The Case

for a

Living Universe

Is there mind in all matter?

Matthew R. Benton

Edited by James Kingsland

Postbridge Books

For all my friends and family.
For Frankie.

With thanks to James Kingsland for his skill and dedication in editing this book. I would also like to mention Victoria Saunders and Chris Garton for their thoughts. Special thanks are due to Tim Lartique, for his invaluable feedback as the all-important first reader.

Contents

Preface

For many years I had intended to write a book containing these ideas. *The Case for a Living Universe* gives my reasons for regarding mind, matter and energy as essentially the same stuff. I believe the main reason this idea seems strange to many is that following the Enlightenment, Western culture created an arbitrary division between our minds and those of other species. This philosophical division was then extended to the rest of nature, and so largely excluded the possibility of mind in the wider material world.

I began writing this book in October 2020, when it became clear that winter in the UK would be dominated by continuing Covid lock-downs. Faced with many months working from home, exclusively in the company of my lovely fluffy grey Nebelung cat, Frankie (who sadly is no longer with me), I had reason to finally marshal the wide range of sources I believed would be relevant into an accessible and hopefully entertaining book. As a friend of mine once astutely observed, "You do projects. That's what you do."

I imagined eight to ten chapters, each having a distinct subject area, drawing on a range of evidence from science, history and culture. Yet when I tried to organise all these disparate sources I found most points would fit in more than one place. For example, did a recent Chinese study, suggesting monkeys have a conscious and a subconscious visual field, belong in the chapter on animal cognition, the one on neuroscience, or a later chapter about how Western culture separates us from nature?

The solution to this problem was a pack of coloured sticky-notes from the local supermarket, a quarter each of pale yellow, deep fuchsia pink, then blue and green of a similar tone. On the plain white wall of the back bedroom, the bold pink stood out as 9 chapter headings. As the blues and greens were equally appealing, they held the points I thought relevant. Writing "Monkeys: Conscious and sub-conscious vision" onto a green note forced me to give this piece of information a home under one of the chapter headings.

Those sticky-notes made the planning far more straightforward than I had anticipated. Later I even found a use for the boring yellow notes, where I put questions I needed to answer as I did further research. Disliking these jaundiced-looking notes gave me an incentive to get the task done quickly, crush them into a satisfying ball, and drop them into the rubbish bin.

At some point a blue or green note would have carried these thoughts: "Qualia, quales. Explain"; "Phenomenologists, Husserl etc..." "Quantitative vs qualitative"; "Ontology and epistemology". However, these terms do not appear in this book. I planned the body of the text to be around two hundred pages, feeling that should be sufficient. As space was limited, I also took out those concepts not directly relevant to the case I was making, as I wanted to avoid terminology that would be unfamiliar to a significant number of readers. Otherwise I would have to describe terms specific to the philosophy of mind, and remind the reader of those explanations when they returned later in the text. (In case you're wondering, "qualia" are the much discussed units of conscious experience.)

This book argues that there are several reasons to believe that intelligent awareness, essentially some sort of *mind*, may be more prevalent in the universe then previously understood. However, I am not trying to pinpoint the exact nature of the human conscious experience, so for my purposes, "subjective experience", or "first-person experience" serves well enough for qualia. Hopefully, this approach allows the reader to follow the course of the book without too many interruptions.

Regular readers of philosophy may feel this book lacks other essentials. Philosophy books sometimes contain passages like these: *Let P equal... therefore N... solve for... and thus we find....* Greek letters or logic symbols make an appearance on the basis that sets of statements can formulate an argument, which can be worked through to a conclusion, representing a sort of truth.

The use of logical notation can create a barrier for those not actively engaged in academic philosophy. I am also sceptical about whether this form of debate can ever really lead to any additional knowledge. Strictly speaking, this book is neither a work of philosophy, nor of popular science, although it contains a small amount of the first, and a large amount of the second. I admit to having some sympathy with these thoughts on philosophy from American physicist Neil DeGrasse Tyson in his book, *Letters from an Astrophysicist:*

> I have never been a big fan of philosophy as applied to the physical sciences in the 20th (+21st) century. I have found common arguments to be based more on word usage and word meaning rather than ideas, and so have found the discussions to be largely useless to the progress of science.[1]

On the other hand, scientists often seem to regard information as inherently truthful, without necessarily considering how their philosophical assumptions may affect their conclusions, as surely as a poorly curated data set.

I am also mindful that the scientific evidence presented here, relating to animal cognition in particular, will inevitably be incomplete, and is destined to become less remarkable in a short space of time. For example, writing this preface two years after I began putting notes onto the guest room wall, a University of Tübingen study showed that crows can understand recursive patterns. These patterns are the equivalent of humans understanding a sentence such as "The cat the dog chased meowed". The study has not made it into the body of the text, because I had to draw a line somewhere, even though it was another striking example of a skill that scientists, philosophers, anthropologists and psychologists long argued was only possible with a human brain.

Although this book frequently returns to the artificial divisions between us and other species, it was never intended to be a full-throated repudiation of the myth of human exceptionalism, which has in any case, pretty much had its day in Western culture. This book really has a single point—it appears that stuff happens in nature which may be seen as an element of mind where we would not normally expect mind to be. Even if you disagree with what follows, I hope you will find ideas and information that hold your interest through until the end.

A couple of stylistic points. First, I frequently write "mind, matter and energy" to suggest the mental and physical are one and the same. In physics, matter and energy are considered to be the same—on a continuum, but in a different state. Although I could have chosen matter or energy, or referred to the "the mental and the physical" instead, I believe it simply *reads* better to use that three-word phrase.

Second, I have sparingly used subtitles for sections throughout this text. Most chapters have none, some have several. I could have found subtitles for all the sections in this book. However, I believe this creates a tendency for the reader to read for the next subtitle, rather than concentrating on what the author is trying to say (I myself am particularly guilty of this as a reader). Where I have included subtitles, it is to help structure the chapter, particularly where a large amount of scientific evidence is being described.

Finally, this book is not as complete as I would like. Unfortunately, it is becoming common for traditional book publishers and media companies to charge for the use of even the smallest quotations—a €1,000 flat fee for 33 words in one case. Publishers of academic papers are equally punitive. In my opinion, short quotes are free advertising, and publishers would attract more sales by stating a fair-use word limit as standard. While I have cleared the quotations used here with the relevant publishers, ideally I would have preferred to have used more quotes from modern sources.

1

Alive*ness*

If you were searching the temperate forests of northern Europe or Japan for intelligent life, it is worth noting that one very ancient, yet clever form of life, will be largely out of sight.

Look down from the canopy, where the birds sing in defence of their territory, or work constantly to feed their young. Disregard the lower branches on which clever mammals like squirrels bound and hustle. Ignore the rodents nipping in and out of cover, or even the occasional bear shuffling through the fallen leaves. Forget too about the teeming insect life that animates the forest floor, and ensures that nothing it produces ever goes to waste.

There is an intelligent life-form that thrives in the damp and shady spots, attaching itself to fallen branches and organic leftovers. It is an unappetising yellow ooze called Physarum polycephalum, the "Sponge Bob" slime mould. While it is easily overlooked in its natural environment, given space, a slime mould colony can spread to a metre in diameter in a lab, before drying out to resemble the veiny leftovers of an omelette stuck to a frying pan. In this dehydrated state, it can remain dormant for up to two years, then spring back to life with the addition of a little water.

Until the 21st century, slime mould would only have been of interest to the most committed botanists, those dedicated souls who had found their niche studying the most primitive forms of life. And primitive is most definitely the word here.

Slime mould appeared some 600 million years ago,[2] in the period just before high atmospheric oxygen levels led to the Cambrian explosion, and life finally made its way out of the oceans and onto the land.

Given its early appearance in the evolutionary story, long before the complex neural scaffolding of the human brain emerged, when a blob of slime mould makes its way along a rotting log, its success or otherwise in finding suitable conditions to thrive might appear to be only a matter of chance. We certainly would not expect there to be any real *agency* or intention on its part.

Whether taken as individuals, or a colony of unicellular organisms, slime mould lacks any neurons or anything we might recognise as a brain. It has no central nervous system or sensory organs. However, since the early 2000s, studies have shown that slime mould is remarkably smart. Its abilities include: navigating mazes; weighing up its options; a capacity to learn and pass on that learning; an awareness of time; and being very finicky about its diet. Although I will describe these abilities in more detail in Chapter 3, it is worth giving a mention here to slime mould's party trick—mapping out the Tokyo area transport network more quickly than a team of experienced urban planners!

Slime mould is not the only non-animal intelligence that has been impressing scientists lately. Evidence shows that trees thrive when supported by a healthy layer of mycorrhizal fungi, which take in carbon from plant roots in exchange for phosphorus, forming a relationship that is perhaps more akin to free-market trading than symbiotic dependency. We also know that plants send out warnings of danger to others in their vicinity, and are more likely to help plants that are their relatives. In the last decade, a picture has been building up of plants and fungi that are more aware of, and responsive to their conditions than was long assumed.

Even so, these newly understood behaviours are still generally considered to be the product of automated biological algorithms, rather than being a form of intelligence per se. This is partly because plants are not mobile like us, but *sessile*. It is difficult for us to attribute decision making, or consider possible intent, in forms of life that are not free to move. Yet despite having far less complex biology than the majority of plant life, slime mould can move 1cm per day, accelerating to 4cm when hungry. From its movement we know it is exploring its surroundings, weighing up its options, making choices, and sometimes overcoming its apprehensions. In this sense, the slime mould seems to have found reasons to go where it is going.

Slime mould experiments are part of a growing body of evidence that shows we have greatly underestimated the non-human forms of life with which we share this increasingly wounded planet. While environmental damage, mass poverty and war are by any sensible reckoning humanity's greatest failings, it is also notable that we still lack a secular answer to this basic metaphysical question: our bodies contain the same atoms and energy as the rocks, soil and mountains and nothing more, yet they are apparently aware of nothing. Whereas we are aware of all this, of our own existence, and so much more besides. How is that possible?

I believe the growing evidence of wider awareness and decision making in the simpler forms of life should cause a proper re-assessment of our concept of living intelligence among both philosophers and scientists.

However, when philosophy does address this issue it tends to focus on the neuroscience of human experience, consciousness in the abstract, or the potential for consciousness through artificial intelligence (AI). In truth, it has yet to address the rapidly emerging scientific consensus that the human species' sophisticated awareness is not an evolutionary outlier at all.

To the question of how mind and matter are linked, the mind/body question, there are broadly four possible answers.

These sit on a continuum with mind at one end, and body at the other. There is one philosophical view at either end, and two more in the middle.

One of the answers in the middle is dualism, the idea that the mind and body are separate but complementary things. It is often said that most people are dualist by default, as dualism seems to resonate with our daily lives. We are frequently defined by competing opposites. Do you want tea or coffee? Prefer cats or dogs? Vote conservative or liberal? Arts and sciences are taught in separate classes for separate assessment. We weigh up our long-term interests against more immediate wants. The majority of our cells are replaced every few years, yet we experience life through a constant sense of self that inhabits these ever-changing bodies. There are formless objects in our heads (concepts) and corresponding objects in the real world, which are two different but apparently related things. These separate realms of physical and mental somehow meet in the brain to produce human consciousness.

But when we consider dualism as a description of a more fundamental reality, this supposedly "common sense" view is easily undermined. For example, modern science can detect only matter and energy, not some additional property of mind, so how can we prove mind even *exists*? Dualists also cannot explain where and how these separate spheres would actually meet.

Dualism came to us from Greek philosophers like Plato, then via René Descartes and the Enlightenment. From the Enlightenment onwards, as the physical sciences notched-up more and more wins, some thinkers believed the rational mind was our gift from God, and as such would be the source of human salvation. But as modern science does not need to include God or a soul in its description of what makes us human, first the soul, and then the mind and its subjective experience, was somewhat downgraded in the second half of the 20th century. As a result, the dualism of the Enlightenment has been replaced by the philosophical view behind most modern science: materialism or physicalism.

These belong firmly at the *body* end of the continuum, because they rank mind as secondary to matter, stating that mind cannot exist without the physicality of the matter and energy that make up the rocks, trees, stars, and the rest of the cosmos. On the other hand, according to materialism, matter can most *definitely* exist without mind, and does so throughout the vast majority of the universe. Living beings like us are the rare exception in a universe made almost exclusively of unknowing matter. For the most ardent materialists, the consciousness through which I am writing and you are reading, is simply a by-product of the human brain, like the hum of the cooling fan produced by the computer that sits on my desk.

The third approach is at the other end of the continuum from the *all is matter* approach of materialism and physicalism. It is the largely impractical *all is mind* approach found in the philosophy of idealism. Idealists counter that because we can only know of matter *through* our minds, matter is ultimately dependent on mind, and cannot exist without it. This means that the tree falling in the forest really does not make a sound unless there is someone, or something, around that is capable of perceiving it.

Undoubtedly, most of us have more pressing things to do than explore these abstract questions in depth. Even so, I suspect you would have some idea where you are on the metaphysical scale I have just described.

There is, of course, an alternative to the three philosophical positions outlined above. For as long as I can remember, I have always assumed that whatever enables conscious beings like us to exist is most likely a basic attribute of *all* the matter and energy that surrounds us—meaning that mind, matter and energy really are all the same stuff, none of which should take precedence over the others. Yet I am still surprised how unusual, even exotic, a philosophical position this is considered to be. This is sometimes even the case for those used to dealing with highly abstract concepts, or those well-versed in ancient spiritual traditions. For me it has always been the most common-sense philosophical position of all.

The apparent exoticism of this position may come from its association with religious or mystical experience, represented for example, by the Buddhist saying, *you are the universe experiencing itself.*

I believe another reason awareness as a property of everything seems remote and mysterious, is that the correct philosophical term for what I have described is panpsychism; *pan* giving us everywhere, and *psychism* knowing.

Although there are different flavours of panpsychism, such as panexperientialism and pantheism, which I will touch on in Chapter 5, it essentially comes down to the idea that there is some element of mind in all things. As the Stanford Encyclopedia of Philosophy says:

> Panpsychism is the view that mentality is fundamental and ubiquitous in the natural world.[3]

Sharing this view means I am not dualist, idealist or materialist by default, but panpsychist by default. However, while this book is essentially arguing for a type of panpsychism, I tend to avoid this philosophical label, and will mostly refer to the idea of there being *an element of mind in all things*, instead. In practice this means we live in an intelligent and *aware* universe. But this book is not singing the praises of panpsychism, and there is a good reason for that.

I am fully at home with the idea that every nook and cranny of the universe, everything we know about and everything we have yet to discover, may in some sense be described as potentially aware, intelligent and capable of knowing. Despite this, I have a nagging feeling that the word *panpsychism* will create unhelpful associations for anyone who might otherwise be persuaded that this is a rational position to hold, rather than a belief system for mystics, the chronically naive, and assorted oddballs.

No matter how many times I type those eleven characters in that order, I cannot shake the feeling that the word panpsychism suggests Ouija boards, tree spirits or telekinesis to many, and I am convinced by none of these. While this book is making a case for there being some element of mind in all matter, as a way of accounting for the existence of living consciousness, it does not require the reader to be religious, or to subscribe to any New Age suspensions of disbelief.

So why aren't more thinkers on-board the panpsychic train? As a philosophical idea it has many upsides. For instance, panpsychism quickly dispenses with the mind/body unification problem that plagues dualism. After all, we can account for consciousness in the individual by stating the whole universe is aware, so any point in the universe is also aware and the problem is solved, right? Unlike idealism, it also gives matter its proper status, making it compatible with modern science. Panpsychism is also a plastic enough concept to suit both the devoutly religious and the committed atheist, because it works with or without a deity. It even offers a novel way to account for the counter-intuitive observations found in the bizarre realm of quantum physics, which have long challenged conventional understanding.

Apart from the name, the other major barrier for anyone arguing that all things contain some element of mind is that it could seem to imply that *everything is conscious*—as if we could take the properties of the human mind and scale these down to the smallest level. This is something of a chicken-and-egg problem, because our concept of intelligent consciousness is largely shaped by our concept of the human mind.

For an intelligent, aware universe to be credible, a significant shift in our understanding of what it means to be human, and a recognition that our species may not be evolution's crowning achievement, is needed. This is a hard habit to break, because due to our monotheist religions, and then secular science, we humans have placed ourselves at the top of the pyramid, relegating the rest of nature to being the background scenery for the all-important story of human destiny.

I believe that this subjective, sometimes brilliant, sometimes deeply flawed thing called human consciousness, may just be a subset of awareness in a universe that is at its most basic level, aware, creative and able, in visible ways, to make decisions.

The relative consciousness of species is not then a hierarchy of intelligent and aware beings, with humans at the top, experiencing the most enlightened point of view in the universe. It is a more fluid picture, where the consciousness of species overlaps, and human consciousness is only one *form* of awareness.

But this is not animism or vitalism. Today, no scientist looks for a special substance to breathe life into apparently lifeless atoms, like the spark that animated Frankenstein's monster. The Victorian concept of vitalism is as quaint and ludicrous as a phrenologist devising the profile of a cold-blooded killer by measuring the skulls of inmates in an asylum. However, as artificial intelligence is becoming ever more sophisticated, the question arises whether human inventions might spontaneously become conscious? There are even serious ethical debates about whether robots may deserve rights. These questions arise despite humans being unable to provide a secular explanation for the origin of consciousness in the natural world—a natural world we have been transforming at a rate that is both impressive and alarming.

The fact is, despite science transforming every aspect of our lives, something is still missing from modern science's description of the world.

While it can account for the processes seen in the natural world using evolution, biochemistry and mathematics, it has yet to answer this fundamental question: if *all* is matter and energy, how does a carbon atom from a rock, which finds its way into the food chain and then into the tissue of the human brain, play a part in the conscious experience that you and I are having right now?

This unanswered question of how mind can emerge from energy and matter alone was how the materialist scientist Dr Jim Al-Khalili concluded his fascinating 2008 BBC series *Atom,*[4] which is a reflection of the fact that most scientists believe this problem is still unresolved.

As atoms are un-aware, un-knowing and un-feeling, there is always going to be a problem of *radical emergence*, a question of how and when those atoms manage to switch themselves "on" in our brains to create a conscious human being, capable of experiencing the world? At what point do we draw a distinction between the conscious and the completely unaware? Materialist science seems to allow something (consciousness) to just appear out of nothing (unaware matter), which is a rather unscientific notion.

Strict materialists will, or course, vigorously dispute that there is a radical emergence problem at all, often arguing that the human brain generating consciousness is rather like arranging metal, glass, and gas into the form of a light bulb and adding electricity to produce light. They would say, we just need a bit more information from neuroscience to fill the gaps and understand exactly how this works.

However, that light bulb analogy has limited value, not least because light exists without light bulbs! While a light bulb needs to be constructed properly to produce light, it can only produce particular forms of light, and light bulbs are one of many sources of light found in our universe. In a similar way, maybe it does not require the complex biological machinery of an animal brain to produce intelligent awareness? Maybe awareness, an element of mind, is already built-in and inseparable from matter?

Oddly enough, idealism, the notion that everything is dependent on mind, is perhaps attempting a comeback on the frontiers of science, as I will describe in Chapter 4. This impractical approach seems to be preferred to a philosophy that identifies mind and matter as potentially the same stuff.

It seems the idea that all things contain some element of mind remains in the metaphysical "maybe" pile for now. It is still very much an outlier, because apart from those who are used to debating philosophy of mind in academia, the word panpsychism, and its apparent plasticity, makes it seem esoteric, vague or elusive to many. To others it sounds downright ridiculous.

In arguing there may indeed be an element of mind in all things, this book has three main strands. First, what does the current research on the natural world reveal to be the differences and similarities between our mental lives and those of other species, and what does this tell us about the true status of human consciousness within nature?

Second, what is science finding that might suggest there is mind in more basic biology, and at the smallest scales? And what connection might there be between the sub-atomic world and human consciousness and neurology?

Third, how have philosophy, language and human belief systems created a culture that has separated humans from the rest of the natural world, and through the myth of human *exceptionalism* placed human conscious experience in its own special bubble?

I should state here that, this book does not contain any slam-dunk pieces of evidence for mind in all things. There are no messages of profound insight from ancient texts, or any brilliant examples of deductive reasoning that lead us to an unavoidable conclusion. There are no Zen koans that temporarily suspend the analytic mind, in order to trigger a moment of pure understanding in the reader. Neither is this an especially spiritual book, as it aims to engage the mind more than the spirit. However, it is also important to remember that *awareness everywhere* is a view found in many of the world's oldest spiritual traditions. In this sense, this book is only giving an updated take on a very old and simple notion.

What this book shares with the mystics is a belief that there is a vibrancy and, for want of a better word, an *aliveness* to all things. Aliveness is, of course, different from the biologists' definition of life, which categorises the majority of matter in our universe as not living. But as I will discuss in Chapter 8, this *aliveness* is something mystics often experience. For the mystic, it usually goes with a sense of being liberated from their life story and current identity, and a perception of their existence as a temporary manifestation of life in a fundamentally living universe.

However, this does not mean the reader must seek out mystical experience to grasp the concept of mind in all things. It can certainly be understood through reason, and backed up with empirical evidence.

To this end, this book presents scientific research, which is mostly very recent, and examines the structures of understanding we have inherited through science, philosophy and culture, that tend to make the idea of aliven*ess*—of mind in all things—seem remote, strange or exotic. I am also flagging what I believe to be the inherited biases and blind spots, that for centuries have led us to believe such a thing is either mysterious or impractical. By this I mean, the argument in this book is as much one of subtraction as it is addition.

Topping the list of biases is our tendency to put human experience above the lived experience of all other species, which is a form of hubris—something the human race has an almost fatal tendency towards. It is so much a part of the background radiation of human culture that to even observe it, requires us to step back. This hubris is causing profound environmental damage and threatening our own species' survival, as we fill the oceans with plastic and agricultural run-off, preside over mass extinctions, and fail to get a grip on climate change.

Now don't get me wrong here, humans *are* a unique species with a unique way of understanding the world. It is with good reason we claim to be the most intelligent species. Our ability to shape our environment, to engineer the world to suit our desires, has been the calling card of our species for thousands of years. Changes to our environment have in turn shaped and changed our brains, and fed back into human consciousness. Even if we take my rather pessimistic view of where the human race may now be heading through the misuse of our ingenuity—even allowing for its downsides—the human mind is a pretty good candidate for being in a fundamentally different category from other species. Clearly no other species does what we do.

However, when we try to quantify what makes us special, when we break down those traits and abilities that are considered the key components of that unique human consciousness, it turns out that none of them are uniquely human after all.

Even at the end of the 20th century, anyone claiming in a scientific journal that a species other than humans might be capable of abstract thought or emotion, risked being accused of unscientific anthropomorphising in the following issue's letters page.

But there has been a considerable shift in the scientific consensus about other animals' mental capacities. The sheer volume of research into animal cognition in recent decades shows that traits previously regarded as solely human, such as abstract thought, tool use, planning, altruism, moral behaviour and many more, can be found in other species to some degree.

For example, when given the opportunity to steal a forbidden food treat, dogs appear to be aware that the person nearby might see the theft, suggesting dogs can understand the perspectives of others.[5] Human settlements have shared food and shelter with dogs for thousands of years, so why has the question of whether dogs might be able to anticipate our behaviour, to have some theory of another's mind, only been considered worthy of scientific study well into the 21st century? The notion that dog behaviour is mostly determined by instinctive stimulus and response reactions, which can only be overridden with training, came to Western culture with Pavlov's bells over 100 years ago. It has taken a considerable length of time for us to explore whether that is only one aspect of a dog's mind.

The status of other animals' cognition is now rightfully being re-assessed, and we need to think about what this means for the status of human cognition. What assumptions and biases about the human mind have we inherited through centuries of Western thought that have prevented us appraising the minds of other species, until now? Equally importantly, how have these biases prevented us from understanding our *own* species?

This book uses a great deal of scientific evidence to make its case, but scientists too can be biased by their own cultural assumptions. As an example, and as I will describe in more detail in Chapter 7, until 2010, the accepted figure for the number of neurons in the human brain was 100 billion.

That figure was frequently quoted in scientific papers without citation, despite being based on limited evidence. Since 2010, the accepted number is significantly lower, at 86 billion, thanks to the work of Brazilian biologist Suzana Herculano-Houzel.

The difference is significant, because 100 billion would give the human brain an exceptionally high neuron density among primates, whereas 86 billion is in line with the density of other primate brains.

Our tendency to view the human brain as uniquely, perhaps even *divinely* blessed among primates, helps to explain why this overestimate stood for so long. And here is another example of unconscious bias interfering with reason, which shows even the smartest people are not immune.

In the late 1980s *The Guinness Book of Records* listed Marilyn vos Savant as the person with the world's highest IQ. She also had a column called Ask Marilyn in the American *Parade* magazine. In 1990 she answered a reader's letter about the best strategy for the 1970's game show, *Let's Make a Deal*, known as the Monty Hall problem.

In the game show contestants are trying to win a car. The contestant chooses one of three doors. The car is behind one of the doors, but behind the other two are goats, and the host knows which door is hiding the car. Once the contestant has made their choice, the host opens one of the other doors to reveal... a goat, a wrong answer. Now the contestant is left with two doors and given a choice. Do you stick with your original choice, or pick the other remaining door? Presented with a choice of only two doors, surely it doesn't make any difference if you stick with your first choice or change to the other door, because the odds must be 50/50 whatever you do?

In her answer, vos Savant explained that in fact your odds of winning the car are two in three if you always switch doors, but only one in three if you do not. While this may seem counter intuitive, it is correct. But her answer received a hostile reaction from thousands of readers who insisted she was wrong, and that the odds of winning were really 50/50.

Many of her critics were experts in related disciplines and should have examined her reasoning properly, rather than digging in their heels. Some refused to recognise vos Savant was right until they had played out several rounds for themselves, or saw computer models that proved her correct. A few did later contact her with an apology.[6]

What some very intelligent people failed to consider is that the Monty Hall problem is a game of two rounds. In the first round, three doors give you a one in three chance of winning. By the second round the host has not just randomly taken away one of the three doors, he has taken away an answer that he knows to be wrong, so you are being actively directed *towards* the right answer. If you stick with your original choice you are effectively choosing not to participate in the second round, and your odds must remain at one in three.

The hostile reactions showed how easily emotional attachment and reputation can override the rationality that we humans pride ourselves on. Also, does anyone doubt there was a hefty dose of sexism in some of the criticism vos Savant received? If the same solution had come from a male academic in a university journal, not a woman writing in a popular magazine, I'm sure the reaction from many of her critics would have been different.

What is doubly interesting about the Monty Hall problem is that pigeons seem to understand this problem very well. Chasing a seed reward, not a car, in one 2010 study, after repeated attempts, pigeons recognised the pattern of the outcomes for themselves and switched their second choices at a very high rate, over 96% after 30 days of training.[7] Subsequent studies have shown a lower success rate for pigeons, but most human subjects who were given equivalent training, switched less frequently, and many did not even realise there was a pattern.

Employing remarkable pattern recognition skills, pigeons can navigate huge distances with ease. On this basis it would be easy to dismiss their success with the Monty Hall problem as showing them to be something akin to feathered computers. It is also tempting here to make excuses for human subjects, and suggest their sometimes poor performance was due to feeling intimidated by the unnatural experimental conditions, or that they were distracted, or over-thinking the problem. All of those factors should be taken into account. Yet the truth is, on this specific test of intelligence, pigeons can sometimes outperform people—even professors.

The point of this aside is that human hubris creates blind spots. We just *know* we are right, when we are in fact wrong. Errors become embedded in culture as established fact. Even in the modern age, reason can be greatly affected by context and reputation. We are easily influenced by perceptions of the person delivering the message, which plays into our prejudices.

At this point in human history we have a great opportunity to recognise and filter out human bias, as a result of the broadening out of the concept of intelligence itself. Traditional education systems rewarded the accumulation of facts and knowledge. IQ testing was largely based on knowledge, or the application of reason to abstract problems. Fewer job interviews now include an IQ test, and modern society increasingly values communication and emotional intelligence, such as the ability to read and react to the needs of others.

For anyone who thinks such "soft skills" are ultimately trivial, remember that emotional intelligence relies on what used to be considered a uniquely human ability—the ability to have a theory of mind.

The increasing value placed on emotional intelligence is partly a recognition of the fact that human success has required the well oiled machine of social co-operation as much as it has clever technology. However, emotional intelligence is about more than creating better customer service agents, or more effective managers. There is a more broad and consequential purpose, for all of humanity, to properly understanding our own emotional states. It is the ability to be self-critical and self-aware, in a way that helps us overcome our biases, and the hubris that comes with the simple fact of being human.

Returning to the philosophical problem at hand, a standard objection to there being some aspect of mind in all matter is that inanimate objects like rocks and chairs would need to have minds of their own. Yet, few modern advocates of panpsychism ever argue that rocks and chairs are conscious. In my view, *conscious rocks* is an idea that belongs to animism, not panpsychism, and is a misconception of what *mind in all things* would mean in practice.

The simplest response to any accusations of *rock consciousness* is evolution. Rocks, chairs and other inanimate objects are not subject to Darwinian evolution, and have no reason to be aware of anything. Mental attributes in biological organisms serve the purpose of keeping them alive, allowing them to feed and reproduce.

It seems strange to have to state the obvious, but a chair and a rock are unlikely to be aware entities, because they are not seeking to maintain any sort of existence *as a chair* or *as a rock*. It makes no difference to the rock whether it continues to exist as a rock, or goes through a crusher to become a pile of rock dust.

Similarly, the question of inanimate objects having consciousness is likely to be phrased like this: Do panpsychics think chairs and tables are conscious? People rarely ask if a roof truss, a timber floor or a length of skirting board could be conscious, yet all of these are made from a similar volume of cut wood.

But the chair and table are in our minds conceptually closer to living beings, because they have four legs and are moveable. As with mobile animals vs sessile plants, it is more natural for us to conceive that moveable objects may have some mind, compared to fixed ones.

While a rock may not be able to move itself, as something we might pick up and make use of, our hunter-gatherer brains are inclined to give it an existence as a separate entity. We make it stand out from its environment for our purposes. Some people then employ their imaginations to project mental attributes onto rocks and tables, because those are perceived as separate entities.

The point here is that in suggesting an element of mind could be present everywhere, is not the same as suggesting every object we perceive as a separate entity has an awareness of itself or its environment. Attributing awareness to an entity requires more than simply a perception in the human mind that it is sufficiently separated from its surroundings.

Not only is this book *not* claiming rocks and chairs are aware of themselves, it it also not claiming that *consciousness is everywhere*. As with the word panpsychism, apart from in this chapter, the word *consciousness* does not have an especially high word-count in this book.

In much of what is written about consciousness, there are few attempts to properly define the word, and perhaps it is assumed we share the author's concept of consciousness. Consciousness is often assumed to be about complex mental processing, such as having a strong sense of self, the ability for abstract thought, and so on, instead of the basic evolutionary purpose of enabling behaviours that keep living beings alive.

I am sceptical about how far we can go beyond the observed behaviour of an entity, as you will see in Chapter 6, on neuroscience. Self-awareness, often considered the main indicator of true consciousness, can certainly cloud the issue.

The point here is, anyone who looks for a conscious universe is unlikely to find it, because the word consciousness has many connotations specific to human experience that make a recognition of consciousness elsewhere in the universe, essentially impossible by default. The word consciousness is so subjective and has such a wide range of possible interpretations that it should come with an ever-present disclaimer: *Consciousness.... whatever that is.*

As a result I find it more helpful to focus on the idea of mind, and the external behaviours that might arise from mind, than debates about consciousness. Ask instead, if there were an element of mind in all things, how would that manifest itself? Would there not be mind-like behaviours to observe?

There are two main assumptions in everything that follows. First, that the materialists' radical emergence of consciousness—where brains evolve a certain level of complexity which then generates consciousness—is an unsatisfactory answer to the question of why mental lives like ours exist. I consider that the mystics' saying, *you are the universe experiencing itself* does not mean we are the only points of true sentience in a void of unknowing and unfeeling matter. Rather, it suggests there is some mind permeating all things, and the whole universe may be capable of some form of awareness, knowing and decision making.

Second, in determining whether an entity may be credited with some mind, there are really only three forms of evidence available: setting tests and observing behaviours; comparing biology, especially neurology, with something known to be conscious—which nearly always means measuring it against the human brain; and finally, having a meaningful communication with an entity about its experience. Currently, ours is the only species that can unambiguously provide all three. For many entities, such as the slime mould, we will likely only ever have the first.

Here is another standard objection to panpsychist thought. Some would say taking the behaviour of simple entities as potential evidence of mind is naive animism, because we could apply that principle to a thermostat and wrongly conclude that the thermostat can experience temperature changes—as if the thermostat *feels* hot and *wants* the room cooler, or *feels* cold and *wants* the room warmer. Looking primarily at behaviour in the natural world does not mean thermostats, or any other form of responsive machinery, must then have a mind.

No doubt some will see this as ducking the question, but for me the issue of possible consciousness through AI is clearly a separate one. Thermostats, robots, algorithms or anything else under the category of artificial intelligence, are not products of nature. Those human inventions have a different origin from atoms, bacteria, slime mould, plants, animals, and us.

Evolutionary history matters. Nature has somehow organised itself in such a way to create living beings with minds from basic matter. Following this, humans created machines that may, one day, be determined to have, or not to have minds.

For now, whether thermostats and robots feel or know anything is outside of the scope of this book. How aware, conscious beings managed to evolve from atoms and energy alone is a distinct and different issue from whether artificial intelligence might lead to some synthetic form of consciousness in the future.

Although I briefly mention AI in Chapter 5, this is only to highlight problems with a historical assumption about human consciousness—the idea that it emerges from the human brain's complexity. Instead, what is relevant here is the *behaviour* of energy, atoms, and the biological life that evolved on Earth over several billion years, the overwhelming majority of which happened without any human intervention.

Tracking down from large brained mammals, through the animal kingdom and simpler organic life, we will often only have behaviour to go on. However, the reasons to consider an element of mind may exist in all things are not solely due to these behaviours. They are also based on continuity.

While we cannot be sure whether or not any artificial objects are conscious, we can be certain that evolution has created one form of intelligent, aware life, from apparently unknowing and unfeeling matter—namely us. Not so long ago, it was controversial to consider other animals as anything more than simple stimulus/response machines, without minds of their own.

But as the slime mould experiments have shown, even an organism *without* a brain can be more than a simple stimulus/response machine, and I would suggest it is therefore harder than ever to definitively claim we know the limit of where minds begin in the natural world.

Having set the scope, here is a brief overview of each chapter. The book begins by examining whether human consciousness is really the evolutionary exception, compared with other animals in Chapter 2, then plant life in Chapter 3.

Anyone arguing that mind, matter and energy may be the same stuff must, I believe, make *some* attempt to relate this in practical terms to the behaviours of basic matter. This is the purpose of the thought experiment, and the look at quantum biology found in Chapter 4.

Following this, in Chapter 5 I look at the philosophy of panpsychism itself, and its supposed problems.

Chapter 6 looks at how neuroscience contradicts some of the mythology around human consciousness. Chapter 7 then looks at how our wider culture has reinforced this mythology.

In Chapter 8 I look at whether we might identify a spiritual purpose for mind being in all of nature—although I personally do not believe there is one. In Chapter 9, I conclude by suggesting that ours may ultimately be a creative, decision making universe, with a power and knowledge we might access through technology.

As hinted at in the introduction, this is not intended to be a book that delves into finely argued philosophical points from the history of philosophical debate. I do not have the expertise for that.

Instead, it seeks to highlight that one reason the idea of there being some mind in all of nature seems strange, is the mythology of human exceptionalism that has dominated philosophy, science and Western culture. I have always believed this human exceptionalism to be wrong. Now there is significant evidence that in nature, intelligent awareness is not limited to the human brain.

2

A breed apart?

Here, at the start of the 21st century, there has been a distinct shift away from the idea that human consciousness is the great exception in animal evolution. This insight has not come purely from some innate human quest for knowledge and understanding. It has come, at least in part, from our need to entertain ourselves.

Cute, ever-alert meerkats made great TV in the early 2000s, living out their family dramas in the Kalahari desert in the BBC documentary series, *Meerkat Manor*. The meerkats' entertainment value then gave us reason to test the limits of their cognition. For example, in one episode a meerkat taught her young to disarm a scorpion in three stages by giving the pups a dead scorpion, then a live one made safe, before finally moving them onto the real prey. But with an auditory cue from the researchers, the mother could be tricked into skipping a stage, rather than assessing for herself whether a pup was actually ready.

Cheap video technology has put the natural world under almost as much surveillance as the human animal gets in our city centres. The recording of so much footage from the natural world has unambiguously documented animal behaviours that would previously have been dismissed as observational anecdotes. The evidence is mounting that we are not the only species to go beyond hard-wired behaviours and instincts, something that until very recently was considered up for debate.

The myth of human exceptionalism was never solely due to a lack of data. Human exceptionalism was a philosophical choice being made between the views of Descartes on the one hand (for) and Darwin on the other (against). Western science is the product of a Judaeo-Christian heritage which taught that God has taken us, and only us, beyond the threshold of basic animal awareness, into the realm of moral beings. As the only creatures blessed with souls that would face judgement after death, our existence was the one that really mattered. The fact that at the end of the 20th century, science and philosophy still emphasised differences between humans and other animals, rather than common ground, is a legacy of monotheist religion.

As new evidence has come to light, the idea that the human race occupies a position of wonderful cognitive isolation is no longer scientifically or philosophically tenable. Learning, self-awareness, abstract thought, a capacity for emotions and even language, are being recognised in other species to some extent. It is increasingly clear that minds worthy of the label "conscious", are more widespread in the animal kingdom than previously recognised.

Human control over nature, especially following industrialisation, led to an assumption that by some objective measure, we *must* be the most intelligent species. But was this ever backed-up by evidence? Before looking at the history of how our regard for other species has changed, it is worth spending a couple of pages on an apparently straightforward question: *how should we rank the intelligence of animal species relative to humans?* This is a surprisingly difficult question to answer objectively.

While brain size suggests intelligence, more reliable indicators must take account of brain to body ratios, because larger animals larger nervous systems to control their larger muscles and organs. A horse has a brain about 1/3 the weight of a human brain. But the ratio of brain to spinal cord is 2.5 to 1 in a horse, and about 50 to 1 in humans, suggesting more of the horse's brain will be devoted to motor control than to the kind of processing we associate with higher intelligence. This brain to spinal cord ratio could be used to devise a hierarchy of intelligence. So, for example, bottlenose dolphins are close to humans at 40 to 1. Cats are fairly clever at 5 to 1 and, unsurprisingly, our closest ancestors the apes are more so at 8 to 1.[8]

However, any predictive method for intelligence carries a risk of human bias. Instead of comparing brain to spinal column size, an even more blunt measure is to compare the average brain mass to overall body mass. On this measure humans do not top the table. The average American adult male has a brain about 1.5% of total body mass, which is similar to the ratio found in the guinea pig. However, humans and guinea pigs both score significantly lower than small birds, whose brains may be tiny, but whose bodies are incredibly light.

A more sophisticated variant of this ratio, called the *encephalisation quotient*, was developed in the early 1970s, using a formula to compare expected brain size to body mass, taking account of the characteristics of different species. The environmental requirements of a bird's physical body are significantly different from those of a shark, so it seems reasonable for the ratio to take these differences into account. If we use the encephalisation quotient formula, humans easily top the intelligence table, followed by the bottlenose dolphin and the highest scoring primates.[9] These are the sorts of numbers that fit with our expectations.

But can we independently validate such a formula? Normally, we would take real world data and run it through a new formula to confirm the accuracy of its predictions. As we cannot know how intelligent other species actually are, it is hard to objectively assess whether any new formula is correct. When we crunch the numbers, the assumption that humans are the most intelligent species, means the formula will seem correct if, and only if, humans come out on top.

As animal brains are put together differently, we might look to brain composition to assess intelligence. In humans, the cerebellum mostly handles motor functions, whereas the cerebral cortex is the home of the "higher" functions, such as planning, thinking and language. While an elephant's brain is roughly three times the weight of ours, its composition is very different. Perhaps due to the incredible sensory capabilities of the trunk, 97% of the elephant brain is cerebellum, and only 2% is cerebral cortex, whereas our brains are 75% cerebral cortex. Again, however, composition is not necessarily a reliable indicator of intelligence. For example, other animals with a high percentage of cerebral cortex include the chimpanzee at 73%, and horses at 74%. But which of us would rank a horse as more intelligent than an elephant, despite the elephant's brain having a tiny percentage of cerebral cortex?

Even a larger brain may not mean greater intelligence. Spiders come in a wide range of sizes, from the Patu digua with a body length of 2mm, to the Goliath bird eating spider at 13cm. Despite size differences, studies have shown smaller spiders can be as smart as larger ones. When a spider constructs a web, it makes difficult engineering decisions about how to join the different parts together, leading to occasional construction errors. If bigger brains led to better decision making, we should see fewer mistakes from larger-brained spiders. In fact, large and small spiders make similar numbers of errors.

There is, therefore, unlikely ever to be a method that objectively ranks the intelligence of different species. However, our species has adopted the default position that human minds must be exceptional. We could equally have taken the position that traits such as self-awareness, abstraction and emotion have an evolutionary value, meaning it is unlikely nature would experiment with these traits in only one species. So it is worth understanding why, even in the late

20th century, the mental lives of other animals were considered fundamentally different from ours.

René Descartes was a seminal figure in the development of Western thought. We know he carried out at least one vivisection without anaesthetic, and gave philosophical justification for vivisections to be carried out by others, with wholly untroubled consciences. In his writings Descartes describes vivisecting a live rabbit, and a live dog (although it is not known if he was the one wielding the knife in the second case).

Being clear-minded about the fact that only humans have souls, he was equally clear-minded this meant only humans could really suffer. His description in the late 1640s of animals as biological machines was intellectual justification for many vivisections at the Cartesian Port Royal School in Paris. In 1650, one visitor described routine vivisections of dogs.

The dogs' paws were nailed to a piece of wood, and without anaesthetic they were sliced open to examine blood flow, for as long as they were unlucky enough to remain alive. Being rational men, these Cartesians were confident that the dogs' desperate twitching, writhing, cries and whimpers were nothing more than the reflex reactions of a biological machine, not an indication of actual suffering.

Perhaps we should avoid judging the Cartesians by modern standards. Vivisection was common in early science, and was a research and teaching tool at some of the foundational institutions of modern science, including the Royal Society in London. But we *should* try to understand how reason led people to inflict great suffering on living beings.

To borrow an adage from computer programming: *garbage in = garbage out*. This means the end result is not determined solely by how you process your data, it requires sound data to begin with. It seems Descartes started with an assumption, inherited from religion, that only our species had a soul. He then substituted the soul for the rational mind, which he believed only humans could possess.

A defence for Descartes might be that while he may have taken things a bit far, he was at least putting humanity on the right track, and to be horrified by the suffering caused is to miss the bigger picture of advances brought by the Enlightenment. Personally, however, I do not believe we should write off these vivisections as simply typical for the time, or an unpleasant footnote in the life of an otherwise brilliant man. Had the Cartesians been cutting open these animals to release their demons, or in a quest for the élan vital of living creatures, they would now be dismissed as fools and fanatics from a long distant world of ignorance. Descartes was a leading figure in a movement that has shaped the modern world. But we should consider how pure reason managed to override a natural emotional reaction to what were clearly sentient beings in severe pain.

Although Descartes' view of humans as uniquely blessed came from religion, his view of animals as biological machines was a significant departure from the preceding medieval era, where animals could be held responsible for crimes and even put on trial. Animal defendants could be tried as representatives of a species that had caused extensive damage to crops, or individually for inciting weak-willed humans into performing evil acts. Animals were held criminally responsible and punished like human criminals.

This bizarre fact now seems to us to be proof that the medieval world was dominated by an ignorance and superstition that we left behind, long ago. C.S. Lewis described the rather animistic medieval view of nature, like this:

> In medieval science the fundamental concept was that of certain sympathies, antipathies, and strivings inherent in matter itself. Everything has its right place, its home, the region that suits it, and, if not forcibly restrained, moves thither by a sort of homing instinct.[10]

From the late 17th century, the Enlightenment began the process of sweeping away much of Europe's religious superstition, and in our cultural history its status as a step-up for humanity is rarely questioned. But some falsehoods also became accepted facts in the Enlightenment world, and for the Cartesians, studiously following reason to its conclusion meant wrongly removing mind from all other animals. Medieval courts putting animals on trial was ludicrous, and the punishments for the guilty—hanging or being buried alive —often no better than an animal's suffering on the vivisection table.

However, the medieval belief system did at least credit non-humans with a mind, a mind they believed could have an intent, good or ill, even if all manner of other spirit minds were thrown into the mix besides. Descartes was wrong about other animals' ability to suffer and feel pain. So which is worse? The medieval world overpopulated with minds, or the Enlightenment world where only the human mind was really worthy of the name? In terms of human progress, and specifically our relationship with nature, I would suggest the Enlightenment was not an unqualified step-up, rather a step sideways.

Someone who did not believe our species belonged on a pedestal was the man who gave the world the first credible evolutionary theory, Charles Darwin.

On his return to England after a five-year voyage on the Beagle, and two decades before he completed *On the Origin of Species*, Darwin paid several visits to an orangutan called Jenny at London Zoo. His notebooks show he was impressed by how much Jenny seemed to understand, and he regarded her behaviour, and the mind behind it, as similar to a human child.

He once observed Jenny whining, crying and sulking when her keeper showed her an apple, but would not allow her to have it. After the keeper told her she could have the apple if she behaved herself, Jenny calmed down, was handed the apple, and went off to enjoy it quietly in a chair. She was less forgiving with people other than her keepers however, baring her teeth and hitting Darwin when he tried withholding food. Jenny took up sticks, and even a whip when she encountered giraffes and dogs, apparently knowing these were a more effective defence (or perhaps offence!) than just baring her teeth.

Jenny, and a young male orangutan at the zoo, were also fascinated by a mirror Darwin brought them. In his notebooks, some of Darwin's observations about the orangutans came under the heading "Man", which hints that time spent with them was perhaps contributing to his view that humans had evolved from primates.

The idea that species were not put on Earth in their finished form, as Christianity taught, but developed from other species through the mechanism of evolution, was a key break from religious doctrine. With his theory of evolution, Darwin also saw a physical continuity of species, because with common roots, similar animals would have similar physical attributes. Darwin saw no reason why this continuity should not be applied to mental processes, and believed that humans would be no exception to this principle. In *The Descent of Man* he begins the third chapter by stating:

> My object in this chapter is to shew that there is no fundamental difference between man and the higher mammals in their mental faculties.

He goes on to credit non-human species with significant memory, reason, familial feelings and coordinated action, and draws parallels with human behaviour. Early on, Darwin states that there is no universally accepted definition of mental powers (there still is not) so he must make his case for animal cognition with behavioural observations, some of which came from his own experience.

Darwin noted that his own farm dog would usually attack strangers. But after his five-year absence on HMS Beagle, the old dog came to heel immediately when Darwin walked down the country road to his home, calling its name. It was an example of long term memory in animals, contradicting the idea that animals are simple beasts that live purely in the moment.

In both *The Descent of Man* and *On the Origin of Species*, Darwin includes anecdotal accounts from other naturalists, animal keepers, even game keepers. He repeated a story from a book by a German zoologist about baboons in Abyssinia working together to successfully attack human soldiers by rolling large rocks from the top of a valley to the road below, forcing the authorities to close the road for a week.

Reading *The Descent of Man*, it is striking how much closer Darwin was in 1871 to our current view than what would become the scientific consensus for much of the 20th century. He summarised the chapters on animal cognition by rather sticking a pin in the overinflated human ego:

> Nevertheless the difference in mind between man and the higher animals, great as it is, is one of degree and not of kind.

Because he used anecdotal observations, some argued there was a lack of formal study to back up Darwin's conclusion. It is true that Darwin was making a largely philosophical choice in drawing out the similarities between humans and other animals. Descartes had done the same in setting a clear dividing line between the two. But Darwin's was the more radical conclusion, given that centuries of Western religious thinking still elevated the human experience above all others.

Society was not ready for a wholesale change in our regard for other species, not least because it would have profound ethical implications whilst other animals were treated as an expendable economic resource.

It's worth noting, there are also parallels with the dividing lines drawn between the species and the sexes. A "common sense" view in Victorian times was that men and women occupied *separate spheres*. The male sphere was dominated by logic, reason and worldly achievements, whereas the female sphere was emotional, nurturing, and centred on the family.

These were supposedly complementary spheres, but of course in practice they were on a far from equal footing. In England, before 1918, the fear that a woman's reason might be overwhelmed by her emotional state was used to justify denying women the right to vote. These superstitions lasted well into the late 20th century. For example, the marathon only became an Olympic event for women in 1984, because fears persisted that running long distances would put female athletes in danger of damaging their wombs! Since Victorian times, reason and logic have been regarded as the cornerstones of human progress, and crediting animals with much in the way of awareness or emotion was for many risking a descent into feminine sentimentality, with the potential to drag humanity back to pre-Enlightenment superstition.

Applying Darwin's principle of difference in *extent but not of kind* itself became problematic in the days before science was able to offer much in the way of empirical animal study. While similarities might seem reasonable with the smarter mammals, credulity could be stretched when the principle was applied universally across all species. The Canadian evolutionary biologist, George Romanes, who had been Darwin's friend and research assistant, was heavily criticised for anthropomorphism. He was perhaps himself a little uncomfortable with a strict application of Darwin's principle, when he wrote this in 1883:

If we observe an ant or a bee apparently exhibiting sympathy or rage, we must either conclude that some psychological state resembling that of sympathy or rage is present, or else refuse to think about the subject at all; from the observable facts there is no other inference open.[11]

A different approach came in 1894, from the naturalist C. Lloyd Morgan and his terrier, Tony, who played an outsized role in determining how animals would be regarded for the next hundred years. Morgan observed Tony repeatedly trying to open the garden gate, and eventually succeeding. Morgan concluded that while opening a gate appeared to be a cognitive achievement, it had been arrived at through a process of trial and error—a lower order of problem solving than reason.

Morgan was probably right that his terrier had not experienced much in the way of abstraction or moments of insight as he learned to open the gate. However, when this became the default position for understanding animal behaviour, it shifted the consensus back in the direction of Descartes, and away from Darwin.

Morgan argued it was unscientific to credit an animal's successes to more complex mental processes, if a lower-order process, such as trial-and-error or instinct, would serve as an explanation instead, a principle known as *parsimony*. Like Occam's razor, Morgan's Canon aimed to give the simplest, most parsimonious, explanation for animal behaviour. The terms parsimony and parsimonious are still found in modern scientific writing on animal behaviour, which show the author is conforming to Morgan's standard.

When weighing up which approach is more credible, Morgan's or Darwin's, it should be noted that Morgan's canon is backed up by no more evidence from formal study than Darwin's view. What we now know, and what as a species we *could* have learned if we had followed Darwin's lead and actually looked for it, is that many other animals do have some capacity for abstract problem solving.

In 2015, a BBC documentary set a multiple-stage problem for a clever crow, named 007. Entering the space, 007 sees a food reward at the back of a perspex box, and immediately grabs a short stick that is suspended on a string.

But to reach the food he needs a longer stick, trapped in another perspex box. The long stick can only be released by dropping three stones, trapped in cages, onto a spring-loaded ramp. While the short stick cannot reach the food, it can be used to fetch the three stones.

After first attempting to reach the food with the short stick, the crow realises a different approach is needed. He uses his short stick to fetch the first stone, and drops it into the spring-loaded box. He pauses and checks for the long stick, apparently disappointed it hasn't been released.

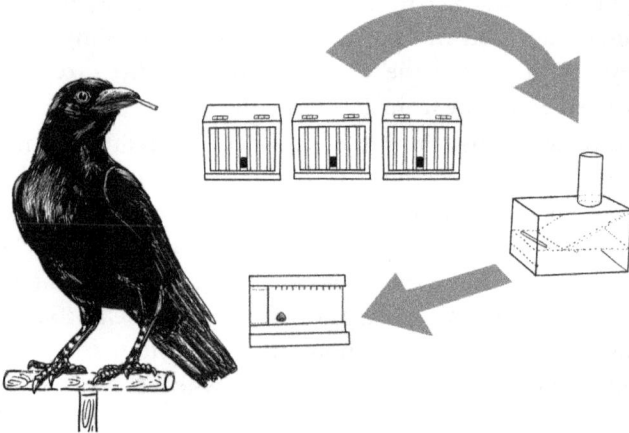

Fig 1. A multi-stage test for a crow.

He then fetches the other two stones from their cages, drops them onto the ramp, frees the long stick, and has his food reward within minutes. What is particularly impressive is that, unlike Tony the terrier, there is very little trial and error in the process. Set a test of similar complication, clearly some humans would struggle to get there so quickly.

The crow had seen parts of this test before, but success with this multiple stage test requires fresh thinking to link the additional steps. Swap perspex for glass, and the technology to create this obstacle course existed hundreds of years ago, even in Descartes' day. In fact, it is perhaps a little patronising of me to call 007 a "clever crow". The evidence from multiple studies is that the ability to problem solve is particularly common among corvids.

An overly *parsimonious* approach to the problem-solving crow would be to cite instinctual behaviour, perhaps look for an inherent tendency in crows to poke their beaks into spaces, which somehow leads to a result through luck or trial and error. Either that or it would be attributed to another low-level, but currently unknown mental capacity. The pursuit of parsimony has often been used to avoid crediting other animals with known human abilities, such as abstraction. But the more animal research is carried out, the harder it becomes to deny the capacity for abstraction in all non-humans. Watch a video of an octopus opening a screw-top jar to get at some crab, then ask yourself, what prey comes with a screw-top lid in an octopus's natural environment? Octopuses can even go one better by opening bottles with *childproof* caps.

Returning to the history of animal cognition, at the start of the 20th century the correct scientific attitude to animal sentience was still up for grabs. Then one famous case, Clever Hans, seemed to bolster the scepticism inherent in Morgan's canon, and expose the risk of anthropomorphising animals by looking for traits that were simply not there.

In 1904, a scientific commission in Germany investigated claims that a horse called Hans could do mathematics. For many years his owner, Wilhelm von Osten, had given shows in which Hans was set simple sums, then amazed audiences as he tapped out the correct answer with his hoof. Some learned people came to believe that Hans was able to do maths.

But the psychologist who led the scientific commission proved that although von Osten was not a fraud, if none of the people around Hans knew the correct answer, the horse got the majority of questions wrong. He realised that Hans was actually reading subtle non-verbal cues from Van Osten or the audience members. People around Hans would tense, changing their posture in expectation as the horse neared the answer. They relaxed when he had tapped out the right number, so Hans stopped tapping his hoof at that point. It turned out that Hans had a talent for reading human non-verbal communication, rather than mathematics.

The high profile case of Clever Hans showed the importance of proper scientific method in animal study, and how easily we can jump to the wrong conclusion. But there is another lesson here. Today Hans would get credit for reading the emotional state of a species other than his own, because empathy and emotional intelligence have greater validity now. This reflects a change in what now counts as intelligent behaviour in our own species, which is more than simply logic, reason and problem solving.

It should be noted that although Hans could not do mathematics, some other animals can. In 2009, researchers in Italy demonstrated that three-day old chicks can do basic sums, including subtraction, without training.

Given a choice, most animals will tend to move towards a larger pile of food—which in itself does not mean that they can count. Instead of food, the researchers used fishing wire to lower small plastic balls behind screens, giving a different total behind each. The chicks watched the plastic balls being lowered out of sight.

When the chicks were released, they went to the screen that hid the larger total, showing they had counted the balls. When the researchers repeated the process, but then picked up and moved some of the balls from the larger pile to the smaller one, the chicks still went to the screen which hid the largest total, showing they had kept track of the changing number, and are capable of addition and subtraction.

There have been other mathematical birds, such as Alex the African grey parrot, who could not only do sums, but even spontaneously introduced the concept of zero into his exercises. This is a significant achievement. It may seem an obvious step to use zero for basic mathematics, and our information age would certainly be impossible without it. Yet zero is not something that emerged organically in all of human civilisation. The Roman numerals employed by Europeans for more than a millennium had no use for zero. Instead the concept of zero was something the Western world learned from India and the Middle East in the 12th century.

Whether or not this was his intention, in the 20th century Morgan's parsimony became the intellectual framework that increasingly placed humans and other species in separate spheres. Morgan did not completely dismiss the possibility of sophisticated human traits, like self-aware consciousness being found in animal minds, but he made these conditional on evidence for such complexity being overwhelming.

This condition is, of course, the catch. Morgan's Canon starts with a baseline that human behaviour is the only behaviour known to stem from true consciousness, putting us immediately into a different class from other animals, unless and until *they* can prove us wrong. This allowed those who followed Morgan to set the bar for evidence as high as they wished, and dismiss suggestions of more complex animal cognition as *Clever Hans* type anthropomorphising.

In the 20th century it became unscientific to emphasise that which we have in common with other species, perhaps because Clever Hans had left egg on some esteemed faces. Instead we could reduce down other animals' sometimes complex behaviours to being the product of much simpler inner worlds than our own. This allowed us to adopt a version of the Enlightenment view that other animals lacked something extra—be it reason, emotion, language or morality, which supposedly only our species possessed.

There were alternatives to a medieval *minds everywhere* position, or the position of Descartes that the human mind is the only true mind. In non-Western cultures where Buddhism, Hinduism and Sikhism dominate, animals were perhaps more likely to be viewed as sentient creatures deserving of compassion, rather than being primarily economic assets.

This is *not* to naively suggest all Asian societies have been exemplars of animal rights throughout human history. After all, a society's norms will always reflect a range of conflicting and contradictory forces. However, when the dominant religion of a culture teaches that animals may be holding a human soul on a journey of re-incarnation, that does tend to lead to more respect for animals, more vegetarianism, and a willingness to give greater validity to an animal's experience.

It should be noted that although Darwin was perhaps Descartes' opposite philosophically, he did not reject all science learned through vivisection, seeing vivisection as sometimes necessary. His attitude to the animals' suffering though was different from that of the Cartesians, as shown by this passage from *The Descent of Man* in 1871:

> In the agony of death a dog has been known to caress his master, and every one has heard of the dog suffering under vivisection, who licked the hand of the operator; this man, unless the operation was fully justified by an increase of our knowledge, or unless he had a heart of stone, must have felt remorse to the last hour of his life.

Having given an overview of how the late 20th century view of animals came into being, I will look briefly at some of the traits long considered to be uniquely human: those of emotion, self-awareness, language and culture. These traits supposedly elevated human consciousness above any other intelligence found in nature.

Animal emotion

One important book that adopted Darwin's approach to animal sentience was *The Question of Animal Awareness: Evolutionary Continuity of Mental Experience* by the zoologist Donald R. Griffin, first published in 1976.

Griffin specialised in bats, and is credited as having proved that bats "see" with sound. In *The Question of Animal Awareness* he presented a range of evidence for more sophisticated animal cognition than permitted by the scientific consensus of the time. Even four decades ago, there was enough evidence from studies of birds, insects and primates to counter assertions from scientists, philosophers and psychologists that other animal behaviour showed no intelligent consciousness.

For example, Griffin pointed out that bees not only communicate the location of pollen through dance, they demonstrate significant memory by waiting until the following day before going to the new location, rather than responding immediately, and have no problem compensating for the changed position of the Sun in the morning. This suggests a type of planning. On learning about new sources of pollen from different dances, bees also weigh up their options, such as distance to the pollen, and make choices that can involve a trade off between quality or quantity.

In Chapter 7 of *The Question of Animal Awareness* Griffin gives a snapshot of contemporary attitudes to animal cognition when he lists the vocabulary for describing animal behaviours, ranging from acceptable to completely taboo, among academics. In the late 1970s he believed it was uncontroversial for a scientist to talk about *neural templates* and *pattern recognition*, yet strictly taboo to use the words *thought*, *choice* or *consciousness*. The second group are words that allow for an animal to have a degree of subjective experience.

The politics of language are ever present when considering animal emotions. When an animal protects its young, can we infer that an emotional inner life is motivating its behaviour, or is it simply evidence of an evolutionary mechanism, which is engaged, almost mechanically, to increase the chances of its genes continuing for another generation? Is it scientific to say an animal appears *content* but not *joyful* or even *happy*? The difference between these words is more than a matter of degree. Semantics has significance and we limit some terms to human experience only, because using human only terms for different species could elevate their experience closer to ours.

Take that specific distinction between being content and happy. Exceptionalists would argue the difference between these words is that true happiness requires a strong sense of who we are now, compared to the past and future, which requires a mental model of the self, relative to the present moment. It is as if happiness only emerges through an awareness of time, tethered to a strong personal identity. Those are certainly traits we can find in humans.

But ask yourself, how many moments of happiness ever came through such an assessment of your entire life circumstances? This sort of reflection may lead to one form of happiness, certainly, but it is far from the only route to feeling happy. Surely, personal history is not essential to a feeling of happiness? We have little issue describing a baby as joyful or happy, despite its having very little self-awareness or personal history. Happiness can be a fleeting emotional state, that can be experienced even when our life is not going as we would like. If human happiness can be temporary, and exist even when our personal histories are exerting little influence on our emotions, why would another animal not be able to experience a similar state?

Evidence for emotion in animals would be suggested by behaviours like empathy and self-sacrifice, and examples do exist in the animal kingdom. In 2013 an underwater photographer observed a pod of orcas that contained a young individual with a damaged dorsal fin and missing pectoral fin, which meant it would struggle to feed itself if left alone. It appeared the youngster waited while the family hunted, then got a fair share of the food. It was being taken care of by family, not left to fend for itself. Another orca pushed her dead calf for 17 days before finally giving him up, suggesting a grief she was not able to overcome.

Altruism between species is more intriguing. Orcas are ferocious hunters and in 2009 a pair of humpback whales were observed intervening as a group of orcas tried to catch and kill a seal they had pursued onto a small ice floe. Not only did they get in the Orcas' way, one humpback rolled onto its back, raising the seal out of the water, and used its pectoral fin to keep the seal there, out of the orcas' reach.

The humpbacks' actions had little obvious benefit for them. In a 2008 article discussing this incident several explanations were offered. It was suggested the behaviour could be a form of "mobbing", such as when different bird species gang up on a bird of prey, for mutual benefit, or that the humpbacks were giving the orcas a warning not to mess with them in future.

Another possible explanation is a well documented tendency in some animals to protect smaller animals of different species, as a sort of misdirected maternal behaviour. There was even a suggestion the humpbacks might have an instinctual need to break-up trouble.[12]

All of these are possible explanations, and without having a dialogue with the humpbacks involved, their true motivation will never be known. However, these are all in the parsimonious category, because the altruism is incidental, seemingly the product of instinctual behaviours, rather than an intentional act. There is, however, the possibility that humpbacks were deliberately being altruistic, knew the seal was about to be torn apart, and took pity.

Other incidents of humpback altruism have been recorded. In 2012 humpbacks were observed trying to prevent a grey whale calf being killed by orcas.

In 2017 marine biologist Nan Hauser was pushed, and put under the pectoral fin of a very insistent humpback whale in the Cook Islands for some 10 minutes, while a 4.5m tiger shark patrolled nearby. Hauser is in no doubt that the humpback was protecting her from the shark.

Although parsimonious accounts are often correct, inter-species altruism is also implied when we drop human exceptionalism in favour of Darwin's *difference of extent, not of type* principle. Even in the space of 15 years, the consensus has changed significantly, and were similar incidents described now, I believe there would be some criticism from fellow scientists if deliberate altruism was *not* offered as a possible explanation.

As with happiness, to deny the possibility of altruistic behaviour in other species implies a very narrow definition of altruism. It would require altruism to be possible only within the moral code of a culture, the most noble expression of the rational mind, like Star Trek's Mr Spock, calmly laying down his own life to save hundreds or thousands of others. However, as with the concept of happiness, when we analyse what a term like altruism means, it fits more than solely human behaviour.

Again evidence suggests that animal emotion is perhaps not so different from ours. Although George Romanes had reservations when he suggested that insects might experience emotional states, he may not have been entirely wide of the mark. Humans self-medicate unhappiness with alcohol, and it seems fruit flies may use the same strategy. A 2012 study found sexually frustrated male fruit flies consistently showed a preference for an alcohol-laced food source, a preference that was four times higher than males who had recently got lucky.[13]

Brain chemistry for reward states in the fruit fly has similarities with that in social mammals, including rats and humans, so perhaps even the humble fruit fly can experience frustration and disappointment? It is of course *possible* that although the brain chemistry is similar, there is no associated emotion in the fruit fly. However, if we accept that our experiences are determined by our neurochemistry, treating our emotions as fundamentally different from other animals when the brain chemistry is similar would require our experience to be transformed in some, as yet unknown, way. Without evidence, this sounds like the myth of human exceptionalism.

The question of animal cruelty

The darker side of other animals' emotional life is the question of whether any of their actions might be cruel. Jane Goodall broke new ground with her studies of a population of chimpanzees in the Gombe National Park in Tanzania, which began in the early 1970s.

Goodall started off believing chimpanzees were nicer versions of us, because they spent most of the day socialising and eating bananas. From 1975, however, researchers observed repeated bullying and apparent enjoyment of violence in what became a brutal four year civil war, as the chimpanzee group split into two factions. Groups of males ganged up to beat rivals to death, and the violence went beyond what was necessary for one chimp to overpower or kill, another.

The levels of violence in the Gombe are rarely observed in other chimpanzee populations, so it is possible that decreasing habitat and the trauma of snaring (one in four of the chimps had been snared in the Gombe at the time) had a negative impact on the group's behaviour. But the capacity for cruelty appears to be there, latent, and it surfaced when circumstances became severe enough.

There are also cases of possible cruelty in animals unaffected by human intervention. A group of orcas has been filmed pursuing a grey whale mother and calf to the point of exhaustion for several hours. The calf was killed but only its tongue and lower jaw were eaten. The grey whale tongue appears to be a killer whale delicacy, because the effort of the hunt is not worth the calories gained.

Given that orcas are large brained social mammals and are capable of forming a support network for a disabled orca calf, I suspect they are aware of the suffering they are causing. Despite this, they ignore the suffering, and their hunt could potentially be described as cruel.

However, I cannot state strongly enough that, as a member of a species that has yet to tire of pre-meditated and organised cruelty to its own and other species on an industrial scale, this is not a moral judgement of the orcas. The question of cruelty arises because if an intelligent animal that is capable of emotion and empathy causes great suffering in ways not necessary for its own survival, as chimpanzees and orcas sometimes appear to do, is this not again a difference of extent, rather than type, when compared to our cruelty?

I would suggest not recognising cruelty in other species is partly because the myth of exceptionalism needs to include some negative traits for humans, along with the positive ones. We share more of our DNA with chimpanzees than any other species, so it is possible that the human capacity for cruelty is in our genes.[14] As such, I find it fitting that Wikipedia records the Gombe Chimpanzee War under the headings "date", "location" and "casualties", with the same format it uses for human wars.

Self-awareness in animals

Some may accept some emotion is possible in animals, yet feel the really clever stuff that we are capable of, such as having a self-aware consciousness, or our capacity for language, still separates us from the other species.

As I said in Chapter 1, I think we should view human consciousness, which certainly comes with a high degree of self-awareness, as one *subset* of animal awareness. We are unlikely ever to define a threshold to cross, a consciousness switch that is on or off, which determines whether an animal is or is not self-aware.

It has long been known that other animals can be fascinated by reflections, without our knowing whether they are actually recognising themselves. In 1970, the mirror test, partly inspired by Darwin's experiences with Jenny the orangutan, was developed to explore self-awareness in other animals. By marking an animal, usually under anaesthetic, with an odourless dye on a part of their bodies only visible with a mirror, we can observe if they use the mirror to examine or remove the mark. If so, it suggests they know they are looking at their own image.

Far more species fail the mirror test than pass. Those that succeed include the known intelligent mammals, chimpanzees, orangutans, orcas, dolphins, one Asian elephant called Happy, and also Eurasian magpies.

What is perhaps more surprising is how many monkey species fail. As primate brains are closest to our own, we might expect to be able to test our way down the hierarchy of primates, until we find the point where a lack of neural sophistication prevents self-recognition, but this does not seem to be the case. If a magpie can get a pass, why not monkeys?

It is possible we have not been testing monkeys correctly, as they may not be curious enough about reflections to go through a process of exploration that might lead to a fresh understanding. There is some evidence for this from a study of rhesus macaques in China, that used laser pointers and a food reward to interest the monkeys in the image in the mirror. After some weeks of this, the monkeys did start using the mirrors to groom and examine themselves.[15]

The mirror test throws up some major anomalies for the idea of a self-awareness hierarchy. The finger-sized bluestreak cleaner wrasse is considered by some scientists to pass the mirror test—which supposedly requires the processing power of a complex mammalian brain. In addition, other animals may be recognising their own image without passing our test because they do not respond to their reflection in ways that we expect.

For example, gorillas used to get a fail, but a famous Gorilla named Koko, who lived most of her life in captivity and away from other gorillas, did seem to understand her reflection. To a gorilla, eye contact is a sign of aggression, so gorillas may be unwilling to investigate a mirror, and begin a process that leads to the realisation that they are looking at their own reflection.

When the mirror test was devised it was probably considered unlikely that animals other than the smartest primates would succeed. Now that some other animals pass, there is instead debate among animal behaviourists about whether mirror recognition should really count as evidence for a self-aware mind at all, or just the ability to discriminate self from non-self. I am not sure how valid this distinction really is. Drawing this distinction may well be humans wanting to move the goalposts, again. After all, the mirror test was for decades considered a strong indicator of self-aware consciousness, that is until animals we deem not cognitively complex enough, like the cleaner wrasse, appear to pass.

Even those animals that do not pass the mirror test may have self-awareness.[16] Dogs do not pass, but this visual test is rather unfair for a species in which vision is not the primary sense. A dog's sense of smell is far more acute than ours. Humans have around 6 million olfactory receptors, whereas dogs can have up to 300 million, which will affect how they understand the world around them. Interestingly, dogs' olfactory processing and visual processing are very closely linked in their brains.[17] In fact, some owners have been surprised to discover their pet has gone blind, as their dog seems to run around, avoiding obstacles. It seems some dogs are effectively *seeing* their surroundings through scent alone.

Unsurprisingly, dogs recognise their own scent, yet I suspect few would take this as a clear indicator of a self-aware consciousness, compared to a chimp grooming itself in a mirror.

Perhaps the dog is only passing a test of discrimination, and is only able to tell its own scent from others. But imagine for a moment that the human race had evolved with very limited eyesight, yet an incredible sense of smell, and we had built our technology, consumer culture and education systems around odours. Imagine a world where we awarded Oscars to perfumers rather than movie directors. In such a world, I suspect we would treat the ability to recognise one's own smell as an important indicator of an infant's brain development towards adult self-awareness.

The mirror test is important for our species, because through a combination of physiology and technology we live in a predominantly visual world. Inevitably, the closer an animal's behaviour is to that of humans, the easier it becomes to credit them as having complex minds.

Language

We humans have large language centres in our brains and highly flexible vocal tracts, and language has been key to our dominance of the planet. Language is a distinctively human trait that is hard to find clear evidence for outside of our species.

Although definitions vary, a common requirement is that true language is evidence for, and therefore requires a brain equipped with, the ability to abstract. But we know corvids have some capacity for abstraction, with or without language. Further, primatologist Frans de Waal did not believe language was necessarily a pre-requisite for thought. In his book *Are We Smart Enough to Know How Smart Animals Are?* he argued that while there is no clear evidence for other species having language as sophisticated as human language, we wrongly assume language is a *requirement* for the sorts of processing we identify as thinking.

Other scientists believe that some forms of animal communication deserve the status of language. For decades it was known that ground squirrels have different warning calls for an eagle high above them, compared to a coyote on the ground, because the difference was audible to the human ear. Ground squirrel *chirrups* can be as short as a tenth of a second, and computer analysis has now shown that these short utterances equate to sentences of this sort: "tall human in green shirt coming towards us" or "domestic dog going away". It turns out the ground squirrels have a distinct word for a dog, another for a wolf, another for a human. They chirrup words describing the colour of the shirts people wear, and whether they are moving towards or away from the group. Even more intriguingly, when researchers walked past the colony carrying boards with drawings of squares, circles and triangles, the ground squirrels started using distinct chirrups, individual words, for each shape. True, the ground squirrels' words may be used to form fairly simple and utilitarian sentences compared with the complexity possible in a human language. Yet the sentences contain distinctive words that could be substituted to convey a very different meaning.

The issue with identifying animal language is that there is no Rosetta stone, no key that allows us to talk with other animals. We can discover that a ground squirrel is likely referring to shapes, animals or colours by putting these objects in front of them. But in the safety of their burrows they may be recounting centuries of shared cultural history to one another, negotiating complex social relationships, or discussing plans for the coming week. Equally they may be constantly chirping "Hi!" to one another. We simply cannot know.

There may also be an inherent difficulty with translating some species' communication, if it serves a different purpose to human language. We believe dolphins are highly intelligent, but their evolutionary success came without the need to manipulate or alter their environment, which was key to our species' survival. They do not need to differentiate between different objects placed in different contexts for different outcomes, as we do. Consequently they may not be transforming sentences as humans do, because they are not transforming their environment.

In the early 1960s NASA part funded the dolphin house project run by John Lilly, in the US Virgin Islands. The aim was to establish interspecies communication, which attracted NASA's interest as a possible model for future extraterrestrial communication. In a large purpose-built two storey house, there were three dolphins, Sissy, Pam and a male called Peter. Researcher Margaret Howe's task was to teach one of the dolphins, Peter, basic English. She spent six days a week with Peter in a partly flooded house, with a desk that hung from the ceiling just out of the water, and a bed above water level so that she could sleep next to him.

One major flaw of the dolphin house experiment was that instead of listening in on natural dolphin communication, Peter was expected to make English sounding words, counting "1... 2... 3..." or saying "ball... triangle... oblong", through his blowhole, which is an unnatural way for dolphins to communicate. Although Peter could slightly transform the sound he produced by flexing the edges of his blowhole, like lips, he could only roughly approximate the English words. Concern also grew that

Peter was mimicking Howe's words, without any evidence that he understood their meaning.

Although the dolphin house started with great optimism, it had a tragic ending. Within a year the funding started to dry up, and in desperation Lilly gave Sissy and Pam LSD, hoping for some sort of breakthrough, an action he later regretted. When the funding ran out, all three dolphins were moved back to a tiny indoor pool in Florida. A few weeks later Peter sank to the bottom of his pool and stopped breathing. Dolphins are conscious breathers. They sleep by resting half their brain at a time, remaining minimally conscious, as they must surface to breathe. For an animal whose every breath is a conscious act, you can make your own judgement about whether Peter had the intention to die, and whether his death qualifies in your mind as suicide.

If they had had laptops, digital recording and computer analysis, the dolphin house project may well have been more productive.

There is an ongoing 30 year research project in the Bahamas with a group of spotted dolphins, who have become familiar with the researchers' presence. To test language ability, the scientists dived close to the dolphins, wearing a box that made specific sounds, effectively a word, in a frequency range comfortable for the dolphins. The divers played a game of passing an object between themselves, using the box to generate the word they wanted the dolphins to associate with the object.[18] The box was linked to a computer, which could detect if the dolphins began using the same sound.

Initially, it seemed the dolphins were failing the test, because the software didn't register them using the word. More analysis of the recordings however showed they were using it, but more creatively than the software was anticipating, for example, by transposing it to a different frequency range.

Some studies have found that pairs of dolphins work together on tasks that require co-operation, such as both pulling on ropes at either end of a food container at the same time, in order to open it.[19] Cetacean communication increases during co-ordinated actions, like hunting, but the vast majority of dolphin communication remains a mystery to us. In addition, sound may not be their only means of communication, and it possible that dolphins can use their powerful sonar to "beam" images to one another.

Yet there is limited evidence for cetacean communication having a practical purpose—for dolphins using sound to communicate more abstract concepts to each other. Worldwide, there are mass killings of dolphins, like the one in Taiji, Japan, where hundreds of dolphins are trapped and slaughtered with long knives. Every year dolphins die in this killing spree, surrounded by pod members and family. If they were able to transmit such information to one another, we might expect warnings about this slaughter to be sent out, and remembered.

It could be that one way a dolphin expresses its intelligence is not repetition but improvisation, a skill we generally recognise as requiring significant creativity and intellect. Perhaps one reason we cannot talk to dolphins on our terms is because their communication is closer to, for example, music composition than a spoken human language? Music has little utilitarian value. Of itself, it doesn't feed us, clothe us or give us shelter. Yet we devote considerable time and effort to creating, consuming and understanding music. Lack of material value does not erode its importance, which has been the case for much of human civilisation. Music training has known benefits on the developing minds of children. And music is not only concerned with expressing emotion—after all, a frequent criticism of improvised jazz is that it is *too intellectual*!

Unlike us, dolphins have little need to manipulate and change their environment in order to survive, so much of their communications may have a different purpose to ours, reducing the odds of our being able to translate their sounds into discreet words or phrases. We might speculate that a significant part of dolphin and whale song exists largely for its own sake. It may involve improvised dialogues, more akin to a contemporary jazz performance than any human language we might one day run through an online translator.

Considering this possibility is not based on some New Age notion that whales and dolphins are gentler and more elevated beings than humans. Our species' survival has been heavily dependent upon our communicating solutions to practical problems, yet throughout history we have also found time to create music and other forms of art. As another large brained social creature engages in complex communications, I am suggesting that perhaps their communication could be more artistic than utilitarian, given they have little need to alter the world around them.

To sum up, the scientific consensus for much of the 20th century was that the human species was the only one capable of an emotional life. Only humans had moral behaviour, could solve problems, plan, go beyond instinct or trial-and-error learning, or had the ability to communicate beyond generalised utterances relating to the immediate moment. Most of all, supposedly only we had the ability to be self-aware, to not just think, but to reflect on those thoughts. The word "conscious" has, until this point in history, effectively been a synonym for the human experience.

There is perhaps something of a metaphor for the 20th century attitude to other species in the great thought experiment of quantum physics, Schrödinger's cat, which explains quantum uncertainty by giving an imaginary cat a 50/50 chance of being poisoned.[20] Doubtless I am not the only person who has wondered whether, in the early 20th century, even a thought experiment was more likely to be considered serious science if it included the possibility of death for a small furry mammal.

Science however does not exist in a bubble. It is a product of its wider cultural context, where scientific attitudes change along with societal ones. The child who is distressed by animal suffering is more likely to become a vet, where some anthropomorphism is acceptable, than a research scientist for a pharmaceutical company, where animal experiments may be a requirement of the job.

Several 20th century scientists were clear in their rejection of the view that animals are fundamentally different from us, notably Donald Griffin, and the primatologists Jane Goodall and Frans de Waal. Their research moved the needle away from Descartes and Morgan, back towards Darwin. As de Waal pointed out, the problem with Morgan's cognitive parsimony is that if we stick to the position that only humans have complex inner lives, this is at odds with *evolutionary* parsimony, because human complexity is then the stand-out evolutionary exception.

The scientific consensus, in the last decade especially, has shifted to an understanding that many aspects of the human experience are less rarefied than the religious and Enlightenment world views would have us believe.

This is an important step in understanding that although the human experience is certainly unique, it is just one of many minds produced by evolution. Recognising that intelligent awareness is more widespread in nature makes it less surprising that a basic element of mind may be an essential part of matter and energy. In the following two chapters I will consider whether intelligent awareness could exist at even smaller scales than the animal brain.

3

Intelligent life without a brain

Nestling in 20 acres of idyllic Kent countryside, Down House was home to Charles Darwin and his wife Emma Wedgwood for 40 years, along with a modest staff, and ten children. Today, visitors can see many of the rooms as the Darwins knew them. We can walk through rooms decorated with bold Victorian wallpapers and fabrics. We can stand at one end of the well-appointed living room, containing Emma's Broadwood grand piano, and shelves of novels by Walter Scott, Jane Austen and Lewis Caroll. From these, the Darwins often read out loud for entertainment, perhaps following one of their competitive nightly games at the backgammon board.

In the study is Darwin's black steel-framed writing chair, fitted with wheels so that he could move rapidly about the room. He would push himself from his desk at the centre of the study, over to his microscope at the ceiling height window where the light was at its best, to his bookshelves, or to the other end of the room, where his collected papers were stowed neatly in a single alcove. Given the size of Down House, the alcove shows a rare regard for an economical use of space. This discipline was something acquired after five years on a ship with 73 other men, because The Beagle was only double the length of the Down House dining room.

No heritage industry visit is complete without a gift shop at its finale. But in my opinion, the gift shop is not the place for the best souvenir on sale at Down House. These are to be found by an honesty box in the room off the lean-to greenhouse, where there are wafer-sized brown envelopes of seeds, suggested donation 50p. These can be taken home and plants grown from seeds collected in the laboratory of an extraordinary naturalist, and one of the few private citizens to ever receive a British state funeral.

Darwin spent the later part of his life studying plants rather than animals, and of the twenty or so books written in his lifetime, five of his last six were on botany. At least 80% of the world's biomass is plant matter, and this apparently passive backdrop to the drama of our lives is not usually considered to have any sophisticated awareness.

But as with animals, we are learning that the no-brained plant world possesses some sophisticated mechanisms and surprising behaviours. The extent of awareness in the simpler forms of life was of interest even to Darwin. *The Power of Movement in Plants* (1880), his penultimate book, co-authored with his son Francis, contains an idea which still sounds speculative nearly a century and a half later. The idea that plants may be said to have brains.

> It is hardly an exaggeration to say that the tip of the radicle thus endowed, and having the power of directing the movements of the adjoining parts, acts like the brain of one of the lower animals; the brain being seated within the anterior end of the body, receiving impressions from the sense-organs, and directing the several movements.

The *endowment* of the radicle (the root) the Darwins were referring to, was its sensory capabilities. They had found that root tips were sensitive to light, gravity and levels of moisture. The Darwins knew plants could distinguish between hard and soft obstacles, and grow around obstacles even before coming into contact with them, findings since backed up by modern experiments. They believed sensory information from the whole plant was sent to the root tip, where a decision was made on how to progress and signalled back to the rest of the plant.

Most of us will have rotated a houseplant to correct it from leaning towards the light coming through a window, without really thinking much about how this correction happens. We might assume new growth heads in one direction, on top of old growth that was heading in a different direction. But in terms of potential processing, it is significant that the correction in fact happens lower down.

By covering canary grass seedlings with small caps, the Darwins found the caps interfered with the plant's sense of direction, as the very tip of the seedling has the job of sensing. The stem bends some way below the tip, directing it back towards the light. The fact that the tip was instigating directional change further down the plant, led the Darwins to consider that the signal might be reaching a sort of root brain, not too dissimilar to the brain of a simple animal. This brain receives a signal from the tip, then sends the signal for a change of direction back up the plant to the parts that are harvesting light. This idea of the root being able to direct change in other parts of the plant became known as Darwin's root-brain hypothesis.

Perhaps because the root-brain hypothesis was contained in a few lines at the end of one book—and some of Darwin's contemporaries had accused him of amateurism in his work on plants in particular—it was largely ignored by botanists for decades.

Victorian objections to passive, sessile vegetation being candidates for brain power came partly from Christianity placing humankind at the top of the hierarchy in nature. But it was also an attitude inherited from classical civilisation. Just as thinking on animal sentience had been influenced by Descartes, the root-brain hypothesis went against the prevailing hierarchy of a natural order that Victorian science had inherited from Aristotle.

Aristotle was as much a scientist as he was a philosopher, wrong about some things, occasionally reprehensible on others. He was wrong in believing that heavy objects fall faster than light ones, reprehensible in believing slavery was part of the natural order. However, Aristotle was also one of the founders of modern natural history, and made many important advances. He dissected birds' eggs at different developmental stages, which led him to conclude that animal organs developed in an order, rather than being miniaturised and present at conception, as most wrongly assumed at the time.

Most significantly, Aristotle created a zoology of around 500 species, making him the first person to systematically categorise species by their similarities and differences, their environment and behaviours, and his zoology was referred to and extended by scholars for centuries after. So the fact that a figure as authoritative as Aristotle had described plants as "mere living" in contrast to animals and humans, set the tone for Western thinking about plant life until the present day.

Aristotle's conclusion about plant life was not unreasonable for someone studying the natural world without microscopes or the benefit of time-lapse photography. After all, even to modern eyes plants don't really *do* much, do they? And whatever it is they might be doing certainly doesn't happen on our time-scales. We can account for plant activity without requiring any sort of plant brain, or plant mind being present, because plants could prosper through basic evolutionary mechanisms without *knowing* anything about it. Given the right soil conditions, and appropriate levels of water and sunlight, the stronger plants of each species will grow, the weaker die. Some species may appear more aggressive, perhaps winding themselves around competitors, or releasing chemicals to keep pests at bay, or enticing pollinators with more alluring chemistry. This simple existence surely means they are relatively simple bio-mechanisms whose influence on the planet is dictated only by fortune? Seeds can be spread by pollinators, wind or rain, or by hitching a ride on an animal's coat to another environment. Surely it is only chance, interacting with the mechanism of evolution, that determines whether or not they thrive?

Plant movement and sensing

The view that sessile plants are the passive backdrop to our existence is perhaps understandable. Yet, while plants cannot pull up their roots and move to a better location when the going gets tough, Darwin knew that they are far from still. His book *The Power of Movement in Plants*, documented just how much they move around, in a circular exploratory motion he named *circumnutating*, which happens throughout the plant's life, both above and below ground. The way a plant moves and how it grows is not random or purposeless, and with modern time-lapse photography it is apparent that some plants move a great deal. Even at the simplest level, there is plenty of movement in the way leaves are tilted throughout the day, to maximise exposure to available sunlight.

Young sunflowers move during the night in order to face the rising Sun the next morning. To find out how they do this, researchers at the University of California set up a growing chamber that delivered an artificial 30-hour day/night cycle. This odd time period confused the sunflowers, but when the lights were timed to a standard 24-hour cycle the sunflowers were pointing the right way, ready for the artificial sunrise, meaning they have a circadian rhythm just as we mammals do. Not only do sunflowers anticipate the appearance of the Sun, in crowded fields their stems may lean a few degrees in alternating patterns, to reduce shading from the plant in front, which does not happen when crop density is lower.

Every gardener knows a pea or bean plant needs a cane or wires to grab onto. The varieties we cultivate have been bred selectively over hundreds of years to suit our purposes. As we are providing them with canes and wires nearby to make them more productive, we apparently do much of the work for them, by engineering the ideal conditions for them to prosper. However, with time-lapse sequences (particularly those produced by the Italian botanist Stefano Mancuso at the University of Florence) it is clear that peas and beans do not simply grow upwards until they are lucky enough to encounter a suitable support.

One of Mancuso's videos shows a single pea plant sprouting in a pot around two feet from the only available support in the room. It grows up for a few inches, then the direction of growth alters towards the support.

As the plant starts to get within reach, its tendril is cast left and right of the support, with a sweep of around a foot either side, like a fisherman casting a line, until it makes contact with the pole and grabs hold. The pea clearly knows that a support is there, and is trying to attach itself.

Another of Mancuso's videos is perhaps even stranger. He positioned a pair of young bean plants in pots about one foot apart with a single pole between them. The sped-up nature of the footage shows each bean casting around, oscillating left and right until one is able to reach the pole. Strangely, after the first bean is successful, the second then changes its direction of movement away from the central pole, apparently searching for an anchorage. It appears to be aware that the pole has been taken and it needs to find another support.

We don't know how pea or bean plants are able to aim themselves towards a pole without eyes or a brain. But Daniel Chamovitz's well regarded book, *What a Plant Knows: A Field Guide to the Senses of your Garden - and Beyond* (2013) makes the analogy between several plant and animal senses—sight, smell, touch, hearing and proprioception (the ability to know where all the various bits of you are).

Plants know the difference between red and blue light, which indicates whether sunrise or sunset is closest, which would count as a basic form of vision in animal species. Darwin found the tip of a canary grass shoot was guiding its growth. It has since been established that some plants have specific photoreceptors in their leaves that help them determine whether it is currently day or night, rather than the tip which tells it where the light is. Arabidopsis (rockcress) has 11 different types of photoreceptors all with distinct purposes. Tomato plants can differentiate between shade due to cloud cover, the shade of a wall, and shade coming from another plant. There are many examples of how adaptive plants are to light, and as light is their main food source it is understandable that they would have evolved a high sensitivity to different light conditions and qualities.

When a pea or bean plant makes contact with a support it will certainly know about it, as plants can be highly sensitive to contact. Vines speed up their growth after coming into contact with a support, in order to wrap themselves around it and get a firm grip. Plant sensing can also exceed human capabilities—a burr cucumber can detect a weight of 0.25 grams, whereas the limit of human touch sensitivity is around 2 grams. Touch is not always welcomed by plants, however, as even the briefest touch in some species may trigger the leaf to curl up and die.

Plants also respond to many airborne chemical signals. In several species, when leaves are damaged by researchers simulating grazing or an insect attack, nearby plants will increase their defences in response. Volatile compounds are released by the injured plant, apparently to warn other parts of the same plant to prepare for an attack. Nearby plants are able to eavesdrop on these signals and gain a head start on their own defence. In one study, in which researchers tore the leaves of cloned sagebrush plants to simulate an insect attack, levels of response to these chemical signals varied depending on how genetically close the sagebrush plants were to each other. This suggests that chemical signalling is more specific than just a catch-all warning from a plant under attack.

In simple evolutionary terms, it is understandable that warnings could become more tailored to warn a plant's own kin than to alert unrelated plants. Again, it's a form of sensing, a form of awareness, but one that could be attributed to an automated mechanism that does not require any form of decision making or experience. However there are other examples of plants working in co-operation that may hint at more complex arrangements.

Canadian scientist Suzanne Simard has been a leader of research into the networks of mutual dependencies of roots and mycorrhizal fungi under the forest floor, which she has termed the "Wood Wide Web", reflecting their connectivity. In the mid-1990s she established that birch and fir trees exchange carbon. While we don't know the purpose of these exchanges, the effect is one of mutual benefit. When the fir is in shade the birch will send up to 10% of its own carbon. In the winter when the birch is without leaves the fir will return the favour. There are also exchanges of water and nitrogen between trees. There is even evidence that antibiotic bacteria can be sent by a healthy tree to an injured tree in need.

Simard also found that mature "hub" trees sent carbon to smaller immature trees below the canopy, where photosynthesis is restricted, and calls these hub trees "mother" trees because they appear to nurture their own genetic relatives. When mapped out as a network these complex connections between trees are strikingly similar to the internet or a communications network.

Both the World Wide Web and Wood Wide Web are designed for resilience, allowing transfer of information or nutrients while minimising single points of failure that might threaten the network as a whole. Knowing how computer networks can be damaged can inform our forest management, because Wood Wide Webs become vulnerable if too many hub trees are removed.

It is Simard's view that the forestry industry makes the mistake of growing single species to maximise yield, and treats non-crop species as competitors that consume resources, rather than collaborators that can share resources and make the whole wood stronger and potentially more productive. Bio-diversity is greatly threatened by industrial agricultural practices worldwide, and a correction is overdue. As Simard has said, "We have to stop treating nature as our shopping mall".[21]

It also turns out that establishing the right price for carbon is not only a human economic mechanism, in response to climate change. There is a subterranean barter system going on between mycorrhizal fungi and plant roots, where the phosphorus that plants need is traded for the carbon the fungi needs. Not only are the fungi exchanging their product, they may also be shopping around for the best deal. They can hold back their phosphorus in a form the plant cannot use, release it to a plant that is getting good levels of sunlight, or withhold it from a shaded plant.

Not long ago, our expectation of such relationships would have been that fungi grow, connect to roots, and the fungi that are lucky enough to connect to a strong plant with plenty of carbon to offer will thrive, meaning the success of the individual plant is simply the luck of the draw.

Although we do not yet know why exchanges do or do not take place, the carbon trading is certainly a more sophisticated mechanism than relying on good fortune. It could be that a threshold of carbon has to be crossed, which the shaded plant is unable to reach, making the mechanism as binary as a thermostat.

Or taking it up a notch in sophistication, perhaps fungi are able to detect the best carbon source in their immediate locale, and direct phosphorus to the best available option. At a more advanced level, the withholding of phosphorus could even be a strategy that changes how much carbon is returned, effectively driving up the price, requiring some memory, an awareness of time, perhaps even planning.

One thing is for certain, wherever on the continuum of awareness this trading relationship actually sits, it again demonstrates that we have long underestimated the awareness and adaptability of the plant world.

Plant agency and learning

Plants pick up an enormous amount of sensory information from their environment. However, taking the step from plants being more mobile and having greater sensory capacity than previously thought, to non-animal life having any sort of a brain, mind or experience, is still controversial.

The credibility of possible plant awareness was greatly damaged by some flawed late 20th century research. For example, one 1960s study which caught the public imagination claimed increased plant growth when plants were exposed to classical music, but failed to control important variables like moisture levels, or the positive effect of music on those looking after the plants.[22] Perhaps this experiment had been inspired by Darwin, who once tried playing his bassoon to his plants. If so, the 1960s team were probably unaware that Darwin later named this his "fool's experiment".

There were more bizarre claims in a book called *The Secret Life of Plants* co-authored by ex-CIA agent Peter Tompkins, published in 1973. Among other things, Tompkins claimed to have telepathic communication with plants, and that by connecting plants to a lie detector, their responses could reveal whether a crime had been committed in front of them. These claims captured the public imagination at the time and made newspaper headlines. Stevie Wonder even wrote the soundtrack to a 1979 documentary based on the book. But it was a book filled with bad science, methodological flaws and unverified claims. Its wacky "Flower Power" (quite literally) tone became a major hurdle for any subsequent researchers wanting to investigate any notion of plant intelligence or sensing over the following decades.

By the early 2000s there had been something of a change brought about by more sober research. A group of six scientists held meetings as the Society for Plant Neurobiology, and wrote a paper in 2006 called *Plant Neurobiology: an Integrated View of Plant Signalling.*[23] They argued that plants need to co-ordinate and respond to significant amounts of information, make choices, often use electrical signals (as animal nerves do), and use the hormone auxin which has similarities to neurotransmitters in animals. Consequently, they argued for a new field of investigation into plants, called plant neurobiology.

Soon after however, a group of 30 or so botanists wrote a strongly worded response, objecting to the word *neurobiology*, which they saw as problematic, as plants don't have neurons. Others disliked what they saw as an attempt to anthropomorphise the plant world. Controversy over the application of words such as *neurology, brain* or *intelligence* to plants has continued. However, one argument in favour of the term neurobiology is that neurons are, on one level, just cells that can produce an electrical signal. And while plants do not have neurons, they have plenty of connected cells that produce small amounts of electricity.

The fight over semantics goes to the heart of what is considered possible within the plant world. Scientists rightly aim to be as specific as possible with their language, to correctly differentiate one thing from another, for the most rigorous interpretation and most accurate predictions. Yet there is also a role for metaphor, as it allows us to think differently about a problem and explore possible connections we were not previously aware of. The issue with illustrative language and metaphor is that one person's metaphor may easily cross another's red line of credibility.

Anthropologist and trained biologist Natasha Myers described the balancing act that plant researchers sometimes have to perform, in the article *Conversations on Plant Sensing (2015).*[24] Some of the scientists she spoke to were aware that certain word choices might undermine good research. This is not a simple case of wanting to keep to the received natural order set by Aristotle. In fact, their objection to using the same descriptive language for plants and humans, in particular any suggestion of plant consciousness, was that it could lead to plants being measured against human criteria. This then fails to recognise the uniqueness of plant capabilities—turning air and sunlight into matter, for example—and makes plants into what Myers calls a "lesser us".

Researchers, especially those looking to establish a career, are more likely to play it safe by choosing the passive over the active voice for plant actions, i.e. "the subject's leaves were curled" not "the subject curled its leaves". The active voice suggests agency, which is controversial with respect to plants.

Every gardener will have at sometime described a plant as being happy or unhappy in a particular location, or stressed by lack of water, etc. Researchers confided to Myers that among themselves they frequently use similarly anthropomorphic language. While describing a protein as "happy" may not be a problem with colleagues in a lab, such metaphorical speech will be filtered out when presenting to a wider audience.

When plants were just part of the scenery, these terms were perhaps non-controversial, as they were assumed to be metaphorical speech. Now, as the question increasingly comes up about the limits of plant awareness, the associated language becomes more subject to scrutiny.

The extent to which metaphorical language is problematic depends on whether the conclusion being drawn is dramatically different from the current consensus. For example, Daniel Chamovitz's book was called *What a Plant Knows*, and he did not feel the need to qualify the word *Knows* by putting it into quotation marks. While he draws parallels with animal senses, the research he described was not something other scientists would baulk at. It is more difficult to accept when the results of experiments might require a more fundamental change in our perception of plant life.

There is ample evidence that plants have sensing capabilities and can be highly responsive, which is the minimum requirement for an entity to be placed on the continuum from basic living awareness, to full subjective consciousness. It is a higher level of sophistication when an organism can locate itself in time, through memory and learning.

It is not controversial to say that a Venus fly trap has some memory, as there is a localised short-term memory in the way it catches its food.

The Venus fly trap only slams the mortuary doors shut on an insect when two hairs on a leaf are stimulated within thirty seconds of each other. The plant has to remember that something of about the right mass came into contact with the first hair, followed by something of about the right mass on a different hair, two events of the right type following each other within a set time period.

But the idea of learning, which would require longer term plant memory, an ability to adapt to those memories, and possibly some sort of location for central control (a plant brain) is far from being universally accepted.

Habituation is one form of learning in animals, and is a significant step-up in awareness from instinctive reactions. Think about working in a large open-plan office with images flashing on and off screens, people moving around, snippets of conversation coming to your attention, phones ringing, the rise and fall of a siren as an ambulance passes along the street below.

We become habituated to all these potential distractions, and relegate this sensory input to the background because we know the majority of it can be safely ignored. However, a fire alarm would instantly get our attention, as we know fire alarms require a response.

Monica Gagliano, of the University of Western Australia, wanted to establish if plants could also become habituated. She used Mimosa pudica plants, known as the Touch-me-not plant, due to its habit of curling up its leaves as a response to touch.

The mimosa has likely evolved this on-demand ability because curled up leaves are potentially less attractive to grazing animals. In research, it is useful to have a reaction that takes seconds, as it occurs on a human timescale that shows a clear connection between stimulus and response. However, mimosas can also only afford to respond in this way so often, as a curled up leaf loses some 40% of its capacity to absorb light. It has therefore evolved to discriminate between different stimuli, rather than closing up every time the wind blows, or a drop of rain briefly weighs down a leaf.

The test Gagliano set her mimosas was an event they would never encounter in the wild. On a rig, the 8cm tall potted mimosas were dropped rapidly, 15cm onto a cushioned base.

The mimosas experienced five drops in a row, seconds apart, each plant undergoing six lots of sixty drops in total in one day. She found they stopped curling their leaves after a small number of drops. They stopped after roughly the first 6 drops of the first session—after 6 of 360 drops.

After being dropped many times the mimosa were then put through a process of *dis-habituation*, by being placed on a vibrating plate for five seconds. The dis-habituation tested their response to a different, but an equally unnatural stimulus, in order to check they hadn't simply run out of energy during the repeated drops.

The dis-habituation did cause a different response. On the vibrating plate the mimosas closed up their leaves each time, suggesting they were not simply tired when they had stopped reacting. This was then followed up by repeating the drop, during which the mimosas did not curl their leaves, demonstrating that these two unnatural situations gained a different response.

Groups of mimosas were then left for intervals of 3, 6 and 30 days. Following each of these intervals, when the plants were put back on the rig and dropped again, they generally showed less inclination to close up their leaves—for example, after the six day interval, they kept their leaves open after only two or three drops, rather than the initial average of six drops. They also re-opened their leaves more quickly than the first few times they had been dropped. This suggests that not only had the mimosas become familiar with being dropped on the day, they had retained some familiarity with this event, and determined it was something they could safely ignore, up to a month later.[25]

Gagliano had problems getting her paper published. When peer reviewed, one major criticism was of the dis-habituation process. It was argued the vibrating plate was not sufficient dis-habituation because it might be a stronger stimulus than the drop, so perhaps tired plants were only reacting to the stronger stimulus, not actually learning anything? Against that tiredness explanation is the fact that the mimosas were reacting less to the drop several days, even a month later, which should have been more than enough time to recover.

This study suggests a form of learning and long-term memory is at least possible in plants. I am not here making the case either way for whether these capabilities indicate the presence of a plant brain, or whether plant experience goes along with them. But as with animal cognition, such research means we should revise what we mean by terms such as awareness or learning to include non-animal life.

Incidentally, just as a definition of consciousness or mind leaves room for ambiguity, there is even ambiguity in the definition of a brain. A brain is generally thought of as a single centralised location for processing information. However, there are animals with multiple brains, such as the octopus which has nine—one in its head, wrapped around its oesophagus, and one in each leg.

We tend to think of there being a one to one relationship between the brain and the consciousness. This may not even be the case with humans, as the human digestive tract in most ways meets the definition of a brain.

There is a field of research named neurogastroenterology, so named because the digestive tract contains 100 million neurons, and performs the complex process of digestion mostly independent from the billions of neurons in your head. Neurogastroenterology is not solely focused on digestive processes however. It is also investigating the role of the gut in psychiatric conditions like anxiety and depression because the digestive system contains 95% of the body's serotonin. It really is valid to talk about a 'gut feeling' because it may be that your emotional state is at times more dependent on the brain in your belly than the one between your ears. Sure, your gut is not going to write your essays, help you assemble some flat pack furniture, or calculate the value of Pi. Even so, our emotional state is a core part of how we react and who we perceive ourselves to be, so it is intriguing how much of this might be determined by what is, in a sense, a satellite brain.

Brains cannot be defined by a simple neuron count. An ant brain possesses only a quarter of a million brain cells compared to the 100 million in the human digestive tract, yet the ant brain is very definitely a brain. Surely, the most important thing when considering what a brain is capable of, is how aware and adaptable an organism can be. What behaviours does a brain allow for, behaviours that might be considered intelligent?

If we were to say plants have brains, or a form of intelligence, the question might arise, why would nature bother giving plants any intelligence if they cannot move?

The answer may be that plant intelligence is required because their lack of mobility means making the most of any available resources, and mounting defences against attack or difficult conditions, specifically because they lack the option to escape. As plants variously make sacrifices for their kin, compete against one another, co-operate and share resources, and are able to distinguish self from not self, what of the thorny question of whether we should describe them as in some way sentient? Do plants have feelings? Should plants have rights?

Daniel Chamovitz's view is that plants are not intelligent, and cannot feel pain or suffer. Plants have incredible sensory abilities, and are extremely sensitive to certain stimuli, such as the lightest pressure. Plants are certainly aware when they are under attack. We know mammals feel pain, and that they have nerves specifically for the experience of pain, which differ from those that reveal the shape or texture of an object. By contrast, we have no evidence for a similar sensory network in plants for pain.

In most respects we cannot draw direct comparisons between plant awareness and other animal intelligences, because the biology is so different. Still, we can compare behaviours when an entity has the option to respond differently to its circumstances. Whether plants can be said to have any sort of brain is debatable. But the slime mould described in the introduction has no brain whatsoever, so it is worth looking in more detail at its capabilities.

A basic test used with early autonomous robots was whether they could navigate around a U-shaped barrier placed in its path, rather than get stuck moving in the same direction. Slime mould has no problem with this test, and can navigate around more complex mazes than this. In complex mazes, it seems to remember the dead ends it has already explored by leaving mucous trails behind.

A key feature of intelligence is the ability to balance different conflicting impulses and make a good choice, rather than repeatedly following one impulse, and the no-brained slime mould appears to pass this apparently high bar. High levels of light are potentially dangerous for slime mould, which could quickly cause it to dehydrate, so when the same food is found in high or low light, it moves towards the food in low light. But if offered some top quality food in high levels of light, and ordinary food in low light levels, the slime mould is often willing to take a gamble on the extra light, providing the food is of high enough quality to justify the risk. The balancing of different impulses—an aversion to light vs an attraction to quality food—suggests there is some kind of processing happening somewhere in the no-brained ooze that enables it to weigh-up its options and choose a course of action. As mentioned in Chapter 2, honeybees have a similar ability to weigh-up conflicting options.

Slime mould also appears to have some sense of time. Again using the dry conditions that are a danger for slime mould, researchers timed a drop in humidity to occur every 30 minutes, and found the slime mould prepared itself for the coming change in conditions, meaning it has an awareness of time that goes beyond the circadian rhythms that can be found in much of the living world.

Significant scepticism remains about whether Monica Gagliano's dropped mimosas were in fact becoming habituated. However if habituation and learning are considered reasonably advanced cognitive functions, which require a significant processing capability, we would not expect a life form as simple as slime mould to be capable of the same.

Yet it appears to be so. As one of a wide range of experiments, Audrey Dussotour's team in Toulouse set up a bridge with a food reward on the other side. The control group of slime mould would cross this bridge in an hour. The group she hoped to train had to cross bridges laden with salt. Like light and drops in humidity, slime mould has an aversion to salt. At first the trained slime mould took around ten hours to cross the salt bridge, but the more frequently the blobs did it, the faster the crossings became. After around five days, they too crossed the bridge in an hour, unperturbed by the presence of salt.

Although the slime mould's maze navigation is impressive, the salt bridge indicates a different type of memory again. In the maze the slime mould leaves itself a trail of pheromone laden mucous to mark where it has been, so we might say that the slime mould is not actually remembering the dead ends it encounters, just moving away from any routes it previously marked with mucous and accidentally revisits. With the salt bridge there is no choice of routes, and any mucous left behind is not affecting future behaviour. Instead, the slime mould has actually discovered, and somehow become accustomed to the fact that the salt it dislikes is not dangerous, and its change of behaviour means it is storing this knowledge. It was also found that by taking a small amount of habituated slime mould, and adding it to some non-habituated slime mould, the whole colony benefitted from this learning, and it crossed the bridge as if it had all been habituated to the salt.

Even though slime mould is a simpler organism than a mimosa plant, slime mould habituation and learning is oddly less controversial than plant habituation from Gagliano's mimosa experiments. This may be partly because Dussotour and her team repeated the salt bridge experiment some 4000 times! Aware that their research would say something quite radical about intelligence and learning, they understandably wanted to remove any question of whether the results were repeatable. It perhaps also speaks to how we interpret intelligent behaviour that, because the slime mould moves like an animal, we can more readily accept it may be capable of learning than a sessile plant. Of the two, the mimosa has the more complex biology, yet a question mark remains over plant habituation. If we conclude slime mould can learn, but mimosas cannot, this would be an instance where the greater cognitive achievement does not depend on more varied, or more recently evolved biology.

In some instances the no-brained slime mould can match and even outperform humans. Japanese researchers placed oat flakes in the locations of cities on a map of the area surrounding Tokyo. In the first few hours the slime mould visited the flakes, apparently logging their location. After initially spreading itself out, within a day it formed tendrils, creating a highly efficient network between the flakes, with striking similarities to the transport network around Tokyo. Although it was not an exact match, where it had done things differently the slime mould had formed routes that were equally efficient.

In a different experiment, slime mould blobs were put in the centre of a circle of eleven foodstuffs, with varying proportions of carbohydrates and proteins, and went towards the most healthy option.[26] Take a moment to picture this. The no-brained slime mould was presented with a wide variety of food options. Not only does the slime mould recognise which food is best suited to its physiology, it will head towards that food, largely ignoring the other options. By contrast, and without trivialising the obesity epidemic, choosing a balanced diet is something humans consistently fail to do, despite us knowing the negative consequences for our health.

Of course, intelligence is about more than the ability to perform well on one test. We have complex inner worlds which do more than help us decide what to eat, and due to the human range of conflicting emotional and rational impulses, we frequently end up eating unhealthy food. But the slime mould deserves credit here, because a key marker of intelligence is whether an animal consistently makes the most advantageous choice—as the slime mould does with its diet— not whether an animal goes through a complicated mental process before making a bad choice, as humans often do.

It is easy to try using human complexity as further evidence of our superiority, instead of conceding that an extremely simple life-form can actually outperform us in certain tests.

The no-brained slime mould is apparently making intelligent decisions, but it is hard to imagine it having a first-person experience anything like ours. So maybe the basic nature of intelligent awareness is not a human-like conscious subjective first-person experience?

One final area to consider is the question of sleep and anaesthesia in the simpler forms of life. In humans sleep is an altered form of consciousness where the mind willingly ignores information from the senses about the present moment. Plants certainly rest, as they become far less active at night. There is one important difference however, as plants manage far better than animals in constant light, where they lack the downtime of sleep. In this sense sleep in plants may exist more to prevent the plant wasting energy moving around in the dark when its major food source, light, is not available. Sleep may not be vital for healthy functioning of the organism, as it is for most animals.

Like all definitions of the natural world, the idea that sleep would help us divide the conscious from the not-conscious is complicated by more knowledge.

It was recently discovered that the tiny freshwater dwelling Hydra, which evolved millions of years before complex animals like us, becomes less active in the dark and can be woken up with light. Despite having no brain and a limited nervous system, its rest meets the minimum requirements for sleep. These were identified in the 1970s by Irene Tobler, who was herself heavily criticised for arguing cockroaches are capable of sleep, as this implied a more multi-layered mental world in insects than contemporary science was willing to accept at the time.[27] Sleep was assumed to serve the purpose of allowing a significant sized brain and nervous system to rest and repair itself. But the Hydra has neither, which means sleep likely existed in simple animals before large brains.

Another curious fact worth noting is that plants can be anaesthetised with the same compounds that anaesthetises humans and animals. Anaesthetics deactivate the Venus Fly trap, and under anaesthetics young pea plants will curl up and stop moving. The reason this happens is unclear. It may mean the inherited trigger for consciousness in humans is a very basic evolutionary mechanism, perhaps something that evolved for a different purpose, perhaps being a simple energy-saving strategy. Perhaps plants responding to anaesthetics indicates that a form of proto-consciousness is present, perhaps not. Yet the fact is, plants can be immobilised by the same compounds that so decisively flip the switch of waking consciousness in humans. We know plants can see, hear, feel, move and sleep. We could even anaesthetise a pair of beans to temporarily stop them competing for a cane.

Does all this mean plants are conscious? Personally I hesitate to apply that word to plants and simple animals. As I said earlier, if we go looking for a *conscious* universe we are unlikely to find one, because the word consciousness carries too much baggage to be applied to the simplest lifeforms. We are unable to separate the concept of consciousness from the nature of human experience.

As mentioned in Chapter 1, there are three ways to infer an animal has some form of intelligent awareness: by setting tests and observing behaviours; comparing biology (especially neurology) with something known to be conscious; by communicating (or miscommunicating) with it about its conscious experience. Ours is the only species that can unambiguously provide all three. Other animals can give us the first two, and non-animal life will likely only ever provide us with the first. So it is reasonable to make a distinction between conscious and non-conscious, while accepting the line between the two is likely rather fuzzy.

However, not describing plants as conscious does not mean subscribing to the myth of human exceptionalism. Again, I say our consciousness is a subset of a wider intelligent awareness in nature, which potentially deserves the label of mind, because no-brained life behaves in ways that even a couple of decades ago would potentially have indicated a fairly sophisticated mind. We now know a very large part of the scenery, the majority of the planet's total biomass is not, as Aristotle said, "merely living". We have perhaps been so wrapped up in our species self-importance and sense of destiny that we have ignored evidence that in fact, the scenery is highly aware. The backdrop is aware of kin, neighbours, food sources and threats, and seems to have ways to utilise this sensory input for its own benefit.

The latest plant research requires us to update our regard for nature. Nature is not the passive background to the stage play of human experience, as it has been regarded for a large part of Western civilisation. Nature is very much active and shaping itself, with or without us. Plants and organisms are acting frequently as individual entities, apparently weighing up options, perhaps even employing goals and strategies. As one professor of ecology has pointed out,[28] the age old philosophical question of whether, when a tree that falls in the forest it makes a sound if no one is around to hear it, neglects the possibility that the forest itself could be the witness.

4

Small science

I regard consciousness as fundamental. I regard matter as derivative from consciousness. We cannot get behind consciousness. Everything that we talk about, everything that we regard as existing, postulates consciousness.

Physicist Max Planck said these words during an interview for *The Observer* newspaper in 1931. Some take this as evidence that the person credited with originating quantum physics was onto something from the outset, something that scientific materialism has since been unable or unwilling to recognise— that consciousness is not just the icing on the cake of a material universe, but central to it.

However, this interview was one of a series of articles on the beliefs and philosophies of leading scientists. Taken in the context of the whole interview, it is clear Planck was only stating a belief, his personal philosophy about the way the universe is arranged. The fact that he said this during an interview suggests it was not a belief he regarded as trivial, but it was perhaps not a hypothesis he expected would ever be tested scientifically.

A philosophical view may help or hinder our understanding of reality, even for the great scientists, who have the same right to beliefs and philosophical views as anyone else. Einstein was famously critical of unpredictability at the smallest level that was being described by quantum mechanics, which he likened to God *playing dice* with the universe. He was apparently also unhappy with the instantaneous, faster than the speed-of-light state changes that came with quantum entanglements, describing them as "spooky".[29] But the decades of experiment that followed have demonstrated that it was the quantum theorists who were right, because what underlies our universe is far more strange and inconsistent than classical physics could have anticipated.

It is tempting to look at great advances in knowledge and try to frame them within a narrative of conflict, setting up Einstein as the brilliant visionary of traditional physics, while others like Planck realised the existing model was incompatible with the building blocks of reality, and began feverishly plotting a revolution. But that would be completely wrong.

Planck's theory was devised to account for the different frequencies of light emitted by glowing bodies such as light bulbs, and it won him the 1918 Nobel prize for physics. Planck created his eponymous law and constant in 1900, and at the time perhaps didn't realise how revolutionary they really were. It was Einstein in 1905 who built on Planck's work by arguing that light travelled in discrete quanta, which is the basis of quantum physics. This idea was met with scepticism by many at that time, because it was such a departure from contemporary physics. In fact, Einstein was one of its main proponents. The wider scepticism in the scientific community had certainly ended by 1918, however, as Planck's Nobel prize was recognition that quantum physics was a valid and extremely important area of enquiry.

Planck initiated a scientific journey that has since revealed that at the base of all matter and energy are some very strange and counter-intuitive effects. Inevitably, the question of whether this strangeness has a deeper meaning comes up among scientists and non-scientists alike. Decades of commentary, especially with the volume of speculations to be found online, presents a problem for anyone wanting to add anything on this subject. I am aware that bringing up quantum physics in a book about the basis of consciousness may be the point at which the scientifically minded will switch off, especially when non-physicists like me are having their two pennies' worth. Because of the strangeness of quantum mechanics, a New Age or religious perspective is frequently applied, unproven theories are invoked, and a black hole of solipsism seems to swallow up any attempt to suggest awareness might have some part in those strange behaviours.

Those who believe mind is an essential part of quantum physics often refer to the idea of an "observer effect", especially in relation to the quantum double-slit experiment. The observer effect seeks to account for the central problem of quantum physics by suggesting that human consciousness may be influencing, or even creating the universe down at the smallest level.

Both the observer effect and the more specific suggestion of this book—that we may be part of an essentially intelligent and aware universe—assume the roots of awareness and consciousness are more pervasive than the current materialist view, namely that consciousness is an emergent property of the structure of the human brain.

If you have come across the observer effect before, you may be expecting this chapter to argue in favour of it. It does not, because what most people mean by an observer effect is I believe essentially a form of the largely impractical philosophy of idealism.

Some may feel this chapter goes out on a limb and spoils an otherwise reasonably engaging book. However, a book that invokes a wide range of science to suggest that there is an aliveness, an element of mind in all things, must surely attempt to find some evidence of that mind in the building blocks of matter and energy.

The challenge of credulity here is as acute as the one that faced Georges Romanes when he reasoned that insects might have emotions. So I hope you can treat my observations on quantum physics as thought experiments, rather than an assertion that this is the way things are. As a thought experiment, we should ask ourselves—if there were some fundamental element of mind all the way down, and given we cannot use communication or neuroscience to establish this, what behaviours might go along with some level of awareness at small scales? In addressing this, I will look at the "measurement problem" and the double-slit experiment.

The measurement problem is key to understanding quantum mechanics, but it is also in some sense the Achilles heel of scientific method. It shows that the universe cannot always be observed with complete objectivity—an inconvenient departure from the way experiments have been carried out for hundreds of years. The quantum double-slit experiment and its interpretations are quite counter-intuitive, and it is easy to lose sight of its real-world results. So first I will briefly describe the experiment, then its interpretations.

The classical double-slit experiment was devised by Thomas Young in 1801. It demonstrated that light travels in waves not straight lines. Position a suitable light source (sunlight in Young's case) behind a shield and allow light through a rectangular shaped vertical slit to a screen beyond. At the other side you will see a rough rectangle of light, with some fading out around the edges because it is not in sharp focus.

Add a second slit close by and you might expect to see two blurry rectangles on the screen. Instead, the light spreads out left and right, forming multiple vertical bands. The bands in the middle are brightest, the bands get weaker nearer the edges, and there are bands of light and dark between them. This pattern occurs because light does not travel in a straight line, but as waves. The two streams of light collide with one another as they head for the screen, generating a distinctive pattern of interference. (fig 2).

If you are having trouble picturing this, think of standing at a lake-shore, and the difference between dropping one pebble or two pebbles into still water at the same time. With one pebble you get waves radiating out from the spot where the pebble entered the water. With two pebbles, you get two sets of waves, and almost immediately, one set of waves will collide with the other, creating interference, and a distinctly different effect than dropping a single pebble into the water.

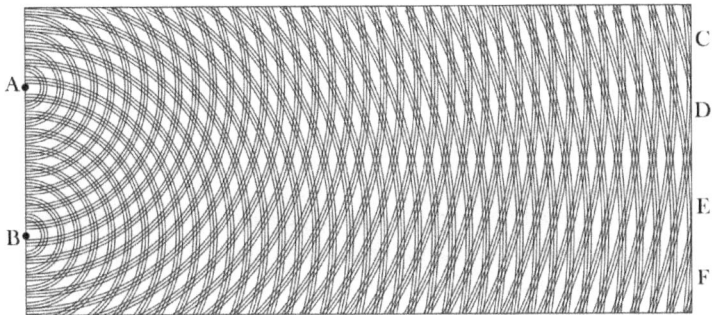

Fig 2. Thomas Young's 1801 experiment. Light waves moving left to right will collide with one another, rather like the waves from two pebbles dropped simultaneously into still water.

From the early 1960s, technology had developed sufficiently to allow physicists to repeat this foundational experiment from classical physics at a subatomic level,[30] and by the early 2000s, much of what we know about the quantum double-slit experiment had been established.

What made these experiments surprising was that at a quantum level the results were not simply a scaled-down version of Young's classical experiment.

To do the quantum version, individual particles such as photons, are released, with one or two slits open. At the other side of the slits, a light sensor logs each photon's arrival and displays it on a computer screen. Fire enough photons at the two slits, a pattern builds up on the monitor, showing where each one has landed.

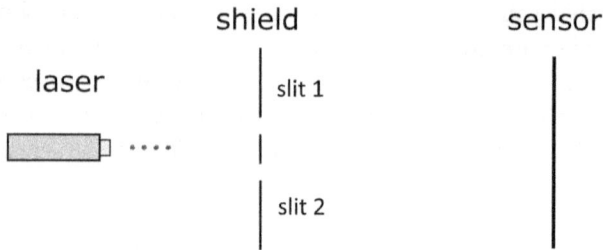

shield   sensor

laser

slit 1

slit 2

Fig 3. Fire multiple photons at two open slits and they will collide to create a pattern of interference.

If you leave both slits open and release multiple photons at the same time, (fig 3) the pattern that builds up on the monitor will be an interference pattern (fig 4). This is to be expected, because even at this small scale, photons are travelling in a waveform, and will collide with each other to create interference.

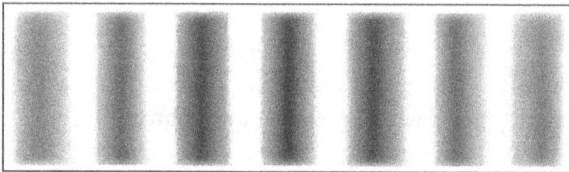

Fig 4. The resulting interference pattern.

If you run the experiment again, but this time block slit 1, so that the photons can only pass through one slit, the photons will form a single blob on the receiving sensor. Clearly, this is because there are no other particles passing through the second slit to create any interference. (Fig 5).

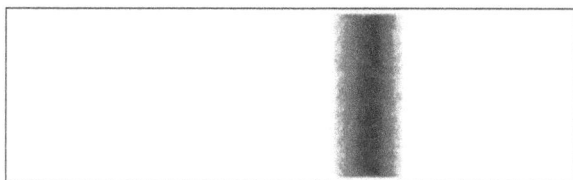

Fig 5. A single blob of photons, without interference.

So far so good. Everything is as the laws of classical physics predict. I should add here there there is the complication of wave particle duality, namely that light can exhibit the properties of both a wave and a particle, a possibility predicted by Einstein and later demonstrated by experiment. However, for the purposes of classical physics, and everyday life, light travels in waves, (as proven by Young's experiment).

Now the quantum weirdness starts. Repeating the experiment with slightly different conditions produces two results that have proven very hard to explain, and sparked decades of debate.

First, if you fire solitary photons through slit 1, with slit 2 left open, because there is no photon travelling through the second slit, you would expect to see a single blob coming up on the monitor. If the quantum world were like the classical world, there should be a blob (made up of dots) that would roughly follow the shape of the rectangular slit the photons were fired through.

But what you actually get is the interference pattern again. (fig 4). This interference happens even though there was nothing present to interfere with those photons as they passed through the slit, one by one. There was no second photon, no second particle stream, and yet those individual photons have behaved as if there were. The solitary photon appears to create its own interference pattern, purely because of the presence of an open slit.

That is what I will call odd effect A.

Understandably, physicists wanted an explanation, so they tried to measure the path of the photon by placing a detector between the slits and the screen. They repeated the conditions that produced odd effect A, by having two slits open and releasing one photon at a time, with a detector positioned to establish which slit the individual photons passed through, before they could reach the other side.

But, strangely, this attempt at measurement changes the result again. The particles go back to landing on the sensor in a blob shape, this time *without* wave pattern interference.

Hence the term *measurement problem*, which I shall call odd effect B.

Odd effect A is strange enough. Odd effect B apparently called into question the basis of hundreds of years of scientific method. Between the Enlightenment and the arrival of quantum physics, we expected the universe to behave like a carefully balanced mechanism, which we could observe and catalogue. We expected our experiments to achieve consistent results, provided we have eliminated all "noise", all unwanted physical influences. Yet, however carefully we try to measure the photon, no matter how objective it seems our measurement *should* be, the process of measurement itself changes the photon's behaviour, and calls into question whether, in the quantum realm, we can ever truly study the world objectively.

That inability to draw a clear line between objective reality and the experimenter, to perform experiments that do not themselves somehow determine the result, has entered popular culture as the idea that there is no such thing as objective reality. That is something of a simplification, certainly. However the fact remains, at a quantum level, reality is not only probabilistic, it is not wholly independent of our actions. The inability to make wholly objective measurements exists, even if you do not believe awareness or consciousness has anything to do with odd effect B.

This weirdness comes with an interesting footnote. Repeating the conditions with the detector left in place but switched *off* actually results in an interference pattern again. No measurement is being taken and the particle's behaviour changes once again.

Now it is reasonable to question whether the presence of the detector, especially a detector that is switched on, is actually exerting a minuscule physical force on the photon that causes it to change its path. But the measurement problem has been with us for decades, and enough very clever people have worked through the problem frequently enough to rule out that possibility. Setting the same conditions but with the detector switched off does not change the particle's path. However, the detector being switched off *does* change whether a genuine act of measurement is taking place.

And that phrase—whether a *genuine act of measurement is taking place*—is where much of the controversy about the quantum double-slit experiment stems from.

You may be familiar with the concept of superposition. Unlike a particle in classical physics, which can have only one location, a quantum particle can occupy many different positions at once.

The standard explanation for this superposition is that before the measurement is taken, the particle can only be described by a wave function, which gives it a probability of being found a particular location. Its position is indeterminate. In this state it can be thought of as taking all possible paths simultaneously through the two slits. Taking a measurement ends this state of probability, causes this waveform to "collapse", and means the particle has a definite state.

Quantum experiments are often conducted under strict conditions, cooled near to absolute zero (-273C), in a vacuum, and using equipment heavily shielded to prevent stray electromagnetic radiation interfering with the experiment—as any energy or matter from the non-quantum world will cause the waveform to collapse. As these points are often missed by people who advocate observer effect interpretations, those who reject the observer effect may claim it is really just a misunderstanding of the measurement problem. They argue that it is not the consciousness of the observer causing the collapse of the waveform, but is simply the result of how quantum uncertainty ends when it comes into contact with the classical world.

There are several interpretations (Copenhagen, multiverse, pilot-wave, etc.) that offer a different take on how we should interpret that waveform collapse, and there isn't space to describe them here. Essentially, the standard view is that the quantum world exists in a probabilistic state, and any interaction between the quantum and non-quantum world (our everyday existence) causes this probabilistic state to become a definite state, and the waveform collapses, regardless of consciousness. As a result, any interpretation that invokes the consciousness of the experimenter is considered by some physicists to be at best a misunderstanding of the process, or at worst, be written-off as pseudo-religious claptrap.

However, it is not clear that consciousness can be completely discounted here. Surely a measurement is only a measurement if someone at some point knows the result of the experiment? This is where the notion of the observer effect enters, because at some point *somebody's* consciousness must be involved. Whether it is the person conducting the experiment, or somebody on the other side of the world looking at an image of the particles hitting the detector a year later, someone must become aware of the results for a complete act of measurement to have taken place. The suggestion is that, it is this *knowing*, the presence of an observer—more specifically the presence of a human consciousness—that causes the collapse of the waveform and triggers the behavioural change at the subatomic level.

There are many ideas about the extent of any observer influence, and some well respected physicists and mathematicians (including Nobel prize winners) have interpreted quantum mechanics in ways that include a role for the observer. A table on Wikipedia showing the main interpretations of quantum mechanics currently lists thirteen interpretations, five of which allow for an observer to have some part in the outcome. In short, among the scientific community, the influence of the observer is a minority, although not entirely fringe view.

One problem with answering this question definitively is that the same experimental outcomes can be taken as evidence by both sides that theirs is the right interpretation. We could try eliminating consciousness by running the experiment, then taking the data that had been logged by a computer, erase the disk, drop it into a blast furnace, or fire it off into space. But what's the use of conducting an experiment if no one learns its outcome? Measurement has to include an act of knowing by someone at some point in time to be of any value. And if you can't take out the act of knowing, how can you be sure whether it is the measuring or the knowing that is causing the change of behaviour?

That is the catch. If you want to find out which slit the particle went through, you need the non-quantum world to detect it. This means it may be the presence of the observer causing the collapse, or the involvement of the classical world causing the collapse, and it is fiendishly difficult to separate the two.

In 2000, a team led by Yoon-Ho Kim at the University of Maryland created the delayed choice quantum eraser experiment,[31] which should have settled this question once and for all. They got around the problem of the non-quantum world influencing the quantum world by having the quantum world itself take the measurement.

The delayed choice experiment makes use of a well established principle in quantum physics, quantum entanglement, which means a pair of entangled particles will reflect the other's behaviour, even at great distance. As with the standard double-slit experiment, single photons are fired from a laser, and pass through one of two possible open slits. However, before reaching a detector, after passing through slit A or B, this single photon is split into two, two entangled photons—a *signal* photon and an *idler* photon.

The entangled pair of photons are then sent on two different paths. The signal photon can only end up at one detector, D1. As with the standard quantum double-slit experiment, either an interference pattern builds up on this detector, or not.

The idler photon, however, goes through a much longer route of prisms and mirrors and will end up at one of four detectors, D2, D3, D4 or D5. Not only does the idler photon take longer to reach its destination than the signal photon, the experiment is set up so that there is a 50% chance the experimenters cannot know which slit the original photon passed through before it was split in two. And there is a 50% chance that this information can be known.

Because the signal and idler photons are entangled, knowing what happened to one photon tells you what must have happened to the other. And because the signal photon has already hit the first detector by the time the idler photon lands on one of the remaining four detectors, all of this convoluted routing should get around the problem of the act of measurement changing the signal photon's behaviour. This is because no measurement is taken until that signal photon has landed.

So what happens when the experiment is run?

Something that leaves us right back where we started. The results essentially match the standard quantum double-slit experiment. Even with that delay in taking the measurement, the interference pattern at the detectors *does change* according to whether the path through the first or second slit was measured.

If the idler photon lands at detectors D4 or D5, the slit is known, the particle is measured, and the signal photon landed *without* wave pattern interference. Whereas, if the idler photon lands at detectors D2 or D3 and does not get measured, the signal photon landed *with* wave pattern interference.

Again, with the delayed choice quantum eraser experiment, differences in interpretation arise. For those who do not believe in an observer role, the result is explained by stating that there is still an act of measurement that ends the probabilistic state.

Others have suggested that particles might even be entangled through time as well as space, that the signal photon somehow communicates its state back in time to the idler photon, so that they can coordinate their behaviours. (It should be noted that, this is a different idea again from the observer effect. The idea of time-travelling particles is itself the subject of several "debunking" articles and videos to be found online.)

On the other side of the fence, others claim the quantum eraser re-iterates that an act of measurement includes an act of knowing, and so because the behaviour of subatomic particles changes, the delayed choice experiment proves there *must be* an observer effect.

I said earlier that I do not really buy into the observer effect, while still arguing it is reasonable to suggest an element of mind is a factor in the measurement problem. My issue with the observer effect is that it is an *anthropocentric* interpretation. I would say that what most people mean by an observer effect, is a form of philosophical idealism, as described in the first chapter.

In idealism, matter is created, or at the very least influenced, by the human mind, rather than the other way round, as materialism states.

For me, to say that the particle's behaviour is changed by the presence of human consciousness, as the observer effect does, is to assert the importance of *human* consciousness in shaping the world. It is as if the wonder of the human mind is not only manifest in its ability to create civilisations, then build nuclear bombs to reduce those civilisations to toxic rubble, it is also capable of influencing matter and energy itself at a quantum level, without us even trying.

That characterisation contains a degree of exaggeration, and I am not suggesting those who argue for the observer effect have some proxy sense of self-importance for the human race. Rather, because our culture contains a level of anthropocentric bias, the idea of a world influenced by the human mind is likely to seem more credible than an aware intelligence manifesting itself down at the smallest level. I suggest this is really no less credible an answer than idealism, but because of our cultural bias, it is harder to conceive of, because unknowing, unfeeling matter is supposed to be a long way down in the hierarchy of awareness, as no more than the background to the human experience.

The other issue with the observer effect is that it is an incomplete explanation. It only addresses the strangeness of odd effect B, described earlier in this chapter—the question of why a particle might change its behaviour when it is measured. It does not offer an explanation for odd effect A, which is equally strange: why would a single particle behave differently when two slits are open, as opposed to when one is open?

Before I continue, again, remember we are in thought experiment mode here, rather than attempting to reach a definite conclusion about the way things are. Given that, let's use our imaginations to consider what the particle is faced with as it approaches the two slits?

We know subatomic particles are behaving differently when presented with different circumstances; one slit or two open; measurement taken or not taken. If we were part of a fundamentally aware and intelligent universe, even at the subatomic level there would need to be *some* capacity to make a decision when options arise. Perhaps the change of behaviour is then the most basic level of awareness and decision making found in nature so far. That offers a possible cause for both odd effects A and B. If awareness is present everywhere, down to every atom and subatomic particle, then perhaps there is enough basic awareness for the possibility of there being another particle passing through the other open slit, to change the first particle's behaviour.

Granted "decision making" may sound too strong a term for simple matter. We are used to thinking of decisions as being only possible in animals with brains capable of weighing up options, and coming to a conclusion, which then leads to a course of action. Even a few decades ago many scientists doubted that any animals other than the most evolved primates, which lacked a sophisticated cortex like ours, could be said to be making decisions at all, instead of just acting through evolved instinctive reactions. But as we have seen, the behaviour of many living things, including plants, suggests many forms of decision making. All mobile organisms make choices of whether to go this way or that way. Other decisions are much more complex because they involve weighing up conflicting priorities. Still, if we were to identify a single *purpose* for animal awareness, including human consciousness, I suspect most of us would say it comes down to the ability to make decisions, according to the circumstances.

For many, the idea of awareness being found in any non-biological matter will be a near impossible philosophical shift. If subatomic particles are behaving in a way that some claim suggests an element of knowing, does that mean they could have personalities, be capricious, or capable of subterfuge? Certainly, once we start considering mind as a possibility at the smallest level, it starts to sound like a medieval animist world view.[32]

However, even as a thought experiment, I would suggest it is unhelpful to think of photons having brains, or being conscious, or anything like. The brain/body pairing is specific to larger animal consciousness, and that word—conscious—is perhaps one that should be reserved for the subjective experience of lifeforms we have reason to believe can have a significant level of self-awareness, or perhaps those that have a distinctive stream of conscious *and* subconscious processing. I very much doubt there is much value in attributing subjective experience to a subatomic particle. In fact it is even slightly problematic to say photons or atoms could be *aware of* the presence of two slits. This may be trying to attribute a sort of bargain basement level of consciousness to atoms, similar to what Natasha Myers called a "lesser us" when plant awareness is compared with ours.

But if all matter and energy had some awareness, it would follow that the whole arrangement of the double-slit experiment, and all the elements in the experiment, are what may be considered aware. What we see in the quantum double-slit experiment is then just a change of behaviour from an entity, a photon, that is currently in a situation where the possible choices being presented influence its behaviour.

Take a step back from the weirdness of decision making being considered at the smallest level for a moment, and think of the strangeness that quantum theory already encompasses: particles can go straight through matter, according to quantum tunnelling; they can move in synchronicity across vast distances with apparently no detectable force connecting them; further, the universe we inhabit is fundamentally probabilistic, or may even be one of a near infinite number of universes. These are all mainstream ideas in quantum physics. There are good reasons to believe in ideas like quantum tunnelling, and as I will describe, these also offer answers for specific problems in biology that have eluded explanation up until now.

On the other hand, it is also worth considering that any explanation of a strange phenomenon at the frontiers of science, is more likely to gain ground if it ties in with contemporary trends.

Integrated information theory (IIT) proposed by the neuroscientist Guilio Tononi in 2004, is a framework for determining to what extent a system may be conscious, by focusing on information as the fundamental unit of the universe. Interestingly, Tononi himself has described IIT as a form of panpsychism.[33]

I make this observation about IIT here without aiming to critique it in any detail. But we should consider whether IIT has more of an appeal than older theories about consciousness because we are in the digital age, where information is key to every aspect of our existence. Perhaps the value we give information here in the 21st century means that we are more inclined to think information itself should be the fundamental constituent of reality, and therefore be the key to understanding consciousness?

Others argue passionately that we are most likely living inside a sophisticated computer simulation. Despite a lack of evidence for, and major problems with, the reasoning behind this *simulation hypothesis*[34] I believe its popularity is partly because so much of what we know and experience is mediated by technology. This can foster a sense of unreality in our daily lives, and a suspicion that we are perhaps being manipulated and controlled. The idea that we are part of a programmed simulation is an extension of this paranoia, which fits with the sense of being somewhat removed from reality, that many of us experience.

Coming back to quantum physics, one of the more popular interpretations is the many worlds or multiverse theory. This explains the appearance of the interference pattern (odd effect A) by claiming that there is an infinite, or near infinite, number of particles from alternate universes passing through both slits at the same time, which collide and interfere with the measured particle, just as if there were a stream of particles passing through both slits.

The theory has been popularised as the idea that ours is one of a near infinite number of universes, existing side by side. We can imagine how differently our lives might have turned out if we had made a different choice at some point in our past—a different partner, career, a whole different life in another country perhaps. Who hasn't wondered how the 20th century would have turned out if someone could travel back in time and kill the young Hitler? We can imagine alternative histories for the human race, as well as our alternative selves, who are able to live out lives that are no longer possible for us, after we take a certain fork in the road. Parallel universes are a staple of science fiction, such as the classic 1967 Star Trek episode, *Mirror, Mirror* where Kirk and his landing party are accidentally beamed onto a vicious dog-eat-dog alternative USS Enterprise.

I suspect the parallel universe idea seems more credible, at least in part, because in modern consumerist society, our lives are filled with constant choices, in which the individual apparently has a high degree of self-determination.

Imagine if we had all of today's technology, but little choice about how to use it. Imagine our work and personal lives being determined at birth, like a medieval serf who must request permission from his master to even leave the village. Would alternative possible lives, and alternative possible worlds in physics, then seem less likely? The point here is that wider acceptance of any radical new scientific hypothesis frequently has a sociological element, playing into our human bias, which is something to be mindful of.

I said earlier that the importance of the observer in quantum mechanics is a minority though not entirely fringe view. Planck seemed to subscribe to it as part of his personal philosophy, as did John Wheeler, who was a major figure in 20th century physics. Wheeler worked on the Manhattan project, coined the term "black hole", developed the idea of wormholes, and taught many important physicists. To establish the extent of observer influence, Wheeler also came up with the thought experiment that later became (in practice) the delayed choice quantum eraser experiment that I described earlier.

In Wheeler's later life, he questioned what part the observer might play in shaping reality, saying for example:

> 'Participator' is the incontrovertible new concept given by quantum mechanics. It strikes down the term 'observer' of classical theory, the man who stands safely behind the thick glass wall and watches what goes on without taking part. It can't be done, quantum mechanics says. Even with the lowly electron one must participate before one can give any meaning whatsoever to its position or its momentum.[35]

More recently, in a 2017 edition of the UK science magazine *New Scientist*, the title of the cover story was "Reality. It's what you make it". The article looked at the work of Markus Müller, specifically his paper "Mind before matter: reversing the arrow of fundamentality". Dr Müller has an unorthodox take on quantum mechanics, that represents a major philosophical departure from the majority of physical science.

He argues that until now science has assumed there was a direction of travel, from the material world into the mind. This traditional view assumes there is an observable world out there, governed by the laws of nature, from which we receive information through our senses. As the basis for centuries of materialist science it has served humans well, allowing us to discard superstition by conducting experiments that tested reality objectively. However, due to the limitations of that paradigm when applied to the quantum world, Müller wants us to consider starting from a point where there are no laws of nature, there is no describable material world, until an observer is introduced. He wants us to consider what happens if we consider the arrow might travel the other way, in a sense putting mind before matter, and then seeing where this approach might lead. His approach is to use mathematical predictions to predict outcomes by focusing on the observer's state, rather than predicting the state of the external world, which the observer then experiences.

Philosophers would of course regard this as a form of idealism. Müller is circumspect about engaging with the wider philosophical implications, and the inferences for our consciousness. I suspect that he has as many reservations about the label *idealism* as I have calling this a book about *panpsychism*—after all, the human mind loves labels. Labelling concepts gives us the impression we can quickly decide which ideas we want to engage with, and which we can discard.

However, even with those caveats, Müller has described his approach as a type of idealism,[36] which makes it distinctly different from the standard materialist approach to reality. Any form of idealism will be at the opposite end of the metaphysical scale from the materialism and physicalism that are the foundations of modern science. Yet a largely idealist approach featured in a leading science magazine in 2017. The point here is, that on the metaphysical scale described in Chapter 1, the idea of mind, matter and energy being the same stuff, is arguably closer to materialism than *any* form of idealism.

When I began writing this book, I searched the UK's New Scientist archive for references to the word "panpsychism", and found only incidental mentions. Repeating the search in 2023, there were some more, since 2020, relating to neuroscience and Tononi's integrated information theory. This suggests there may be a growing interest in this philosophy within the scientific community. However, I wonder if the fact that, some years before this, a largely idealist approach to quantum physics was the cover story in a respected science magazine, is partly because we are more inclined to accept the idea that the *human* mind has the power to shape reality?

Perhaps this chapter's thought experiment, considering how a panpsychist's belief in awareness down at the subatomic level would be visible, has no value to a physicist. To the vast majority of scientists, such questions are philosophical ones, and so not relevant to their work—there's a scientists' maxim that runs, *shut up and calculate*. However, it is interesting that a modified form of idealism may be less controversial to the scientific community than anything that falls within the category of panpsychism. A universe determined by the human mind is, I would suggest, more acceptable because it fits the notion of our being somehow the exception in nature, and our minds occupying the most privileged, enlightened point of view in the universe.

Perhaps a theoretical version of idealism will provide a breakthrough in physics, despite being a historically impractical philosophy. Are the frontiers of science about to be opened up by idealism because it is an idea whose time has finally arrived, or is idealism on the agenda because our species is pre-disposed to believe in the wonder of the human mind? If mind is constantly re-inventing the world around us, is ours the only mind capable of doing this? Idealism was maybe more sustainable when humans appeared to be the only truly conscious beings. But this is no longer the consensus in cognitive science. Does the mind of a chimp or dolphin, a rat or a problem-solving crow possess that same power to shape reality? What about a virus? Or some slime mould that can learn, has likes and dislikes, and can map out the Tokyo area transport network?

Any form of scientific idealism therefore, would need to account for the cut-off point between minds capable of shaping the world, and the matter and energy shaped by those minds.

There is already a large enough scientific problem identifying the threshold at which unconscious matter enables a brain to generate first-person experience—the *radical emergence* of consciousness. There will also be a problem of setting a threshold for minds capable of influencing matter. Remembering these difficulties, another possibility seems equally worthy of consideration—that it is not unreasonable to consider mind, matter and energy as essentially the same stuff.

The above may seem rather a lot to extract from some photons being fired out of a laser in a physics lab. Interestingly though, there is an outstanding question about the barrier between the unpredictable microscopic quantum world and the largely predictable macroscopic classical world that we inhabit. Outside of a laboratory, at what point does one become the other?

Experiments have shown that subatomic particles are not alone in acting differently when tested with the double-slit experiment.

The same experiment has been performed with whole atoms and buckyballs (the soccer-ball shaped molecules made up of 60 carbon atoms). These also create an interference pattern when released individually with a pair of open slits, namely odd effect A.

The same behaviour is observed as we move from individual subatomic particles to the carbon atom. In nature, a carbon atom can find its way from a rock—where it is apparently not aware—into the environment, to become part of a plant, or an animal that a human eats. Through this, it then becomes part of the fabric of the human brain, playing a role in a person being capable of consciousness.

Perhaps this thought experiment—whether the double slit experiment could show mind operating at the smallest level—would seem a little less bizarre if we found similar processes out in the wild, and for this we might look to the relatively new area of quantum biology. Until very recently the quantum world was considered to have little impact on our everyday lives, because quantum effects were so short-lived and almost impossible to view, as particles went from the probabilistic quantum world to the more predictable classical world we inhabit. As noted, it usually requires us to exclude the effects of the classical world, with ultra-low temperatures and heavy shielding, in order to see quantum effects. But this is being challenged with the new field of quantum biology, which suggests quantum mechanics may be essential in many of the most basic biological processes.

One might think that chemistry and biology should, by now, have provided a sufficient explanation for a basic biological function like the sense of smell. But until now they have not. Our best account of smell was that smell receptors come in different shapes, which then fit neatly with the shape of odour molecules entering our nostrils, creating a sort of "lock-and-key" effect. However, the evidence for this is limited, and we only have around 300 different receptor types, while there are over 10,000 distinctly different smells.

An alternative to the lock-and-key theory is "vibration theory". This says, as different molecules have different frequencies, our olfactory receptors may be responding to each molecules' unique frequency vibrations. In practice however, this was even less likely than the lock-and-key idea, because the tiny amounts of energy involved are unable to produce a strong enough effect on our cells.

That was until vibration theory received a boost from quantum biology. What would be impossible for those receptors to do with classical physics would be theoretically possible, if we allowed for our olfactory senses to engage in a form of quantum tunnelling, where particles pass through matter.

Lab experiments using the snappily named technique of "inelastic electron tunnelling spectroscopy" have determined the vibration frequencies of odour molecules, so perhaps our sense of smell works in a similar way? A trial with fruit flies and another with humans appear to show that molecules with the same shape, but modified to have different frequencies, do in fact give rise to the experience of different smells.[37] So the quantum account of the sense of smell is that something quantum is happening, not just under our noses, but *inside* them.

Quantum biology could also account for how birds navigate long distances using the Earth's magnetic fields, how enzymes work, and the causes of genetic mutations, which are the foundation of all evolution. It also has an explanation for one of nature's great mysteries: why a key stage in photosynthesis, where plants turn sunlight, water and carbon dioxide into sugars, is nearly 100% efficient. What follows is a highly simplified description of this process, highlighting why this level of efficiency is so surprising.

Sunlight landing on a leaf causes a magnesium atom to create an *exciton*, effectively a mini battery, which carries its energy to a place called a *reaction centre*, where the plant utilises the energy captured from the Sun. The energy reaching the reaction centre is what allows the chemical transformation of carbon dioxide and water into glucose. However, on the way to the reaction centre the energy of the exciton has to find its way through an array of chlorophyll molecules. These are arranged in such a way that there are many possible routes the exciton could take.

For a simple analogy here, put chemistry aside for a moment. Instead picture a frog in an arcade game, faced with a patchwork of lily pads, which it must hop across to reach the other side of the pond. Before quantum biology, the assumption was that the exciton's energy jumped somewhat randomly between molecules, like a frog hopping on and off lily pads, until it somehow arrived at the other side. The problem is, if the process really was this random, energy would frequently take a much longer route than necessary, with some even getting lost along the way. This would be at odds with a near to 100% efficient process.

The quantum biological answer is therefore, that like the particles in the double-slit experiment—or what happens inside a quantum computer—the incredible efficiency of the process is because the exciton is in a probabilistic state. It is taking all possible routes to the reaction centre at the same time, so it never gets lost, and ends up finding the most efficient route. There is now evidence for quantum processes in photosynthesis, because lab analysis of the exciton appears to show what is called the "quantum beat" of a waveform before it collapses. The quantum beat is considered a signature of a quantum process.[38]

Quantum biology is a new field, so there is understandably caution among scientists before they determine that quantum mechanics is vital to the process of photosynthesis. After all, humans have only been able to reproduce quantum processes under extreme lab conditions, the opposite of what is found in the natural world.

But assuming this theory is right, and photosynthesis involves a quantum process, the question of what behaviours we might expect to find if there were some element of mind in all things, leads to another way to interpret the exciton's action. In an aware, behavioural universe, the exciton would be considered to be aware of, even said to *know*, the quickest route across the chlorophyll—just as the frog knows the shortest route across the lily pads to the other side of the pond.

As with the photon in the double-slit experiment, saying the exciton *knows* is rather like playing pin the tail on the donkey. It is pinning *knowing* onto the exciton, because that is the entity that apparently modifies its behaviour, behaving as if it has some knowledge of what is going on around it. Shifting perspective somewhat, we should instead say the entire system—the exciton, the chlorophyll and the action centre—has an intelligence and awareness that enables the exciton's energy to find its way.

Of course, any idea of knowing at the atomic or subatomic level will sound ridiculous to many, because magnesium atoms are brainless.

But the slime mould also lacks a brain. It lacks even a single neuron, yet it shows some sophisticated behaviours, such as the ability to balance conflicting needs, and an awareness of time. Biologists previously assumed that the exciton travelled unsystematically through the chlorophyll, but this is at odds with the high efficiency of the process. If we were to accept an element of mind in all things, its behaviour would not be simply random. It would have a tendency to head towards the reaction centre, with just enough awareness, enough *knowing*, in the system to allow it to do this.

A thought experiment on quantum processes is not in anyway proof there is awareness, sentience or consciousness down at the subatomic level. For this thought experiment, it is the behaviour that matters. As I said at the end of the last chapter, without our being able to communicate with an entity, or having common biology, our main way of attributing awareness to an entity that we cannot have a conversation with, is to interpret its behaviour. If we repeatedly sent a mouse through a maze, and it consistently behaved differently when there was an open door nearby that another animal might pass through, we would attribute this to the mouse being aware of that possibility. Animals change their behaviour according to their available options, and may also change their behaviour when they are being watched. A change of behaviour because some basic intelligent awareness is present, is the panpsychist way to view both the exciton's and the subatomic particle's actions.

Something that the animals, the slime mould and the photons all share is a degree of unpredictability. We can make predictions about how animals will behave in a controlled set of circumstances, yet sometimes they can surprise us. We also cannot predict exactly *how* the slime mould will explore its surroundings, as it does not simply take the shortest possible route to the oat flake. In the same way, while we can predict an interference pattern will result, as we release photons individually, we cannot know where each individual photon will land.

Of course, depending on where you are on the metaphysical scale I described in Chapter 1, all of this may seem to be comparing apples to oranges, because the mouse has a brain and sense organs that enable it to know there are other options. On the other hand, we can strongly suspect the mouse is aware, even conscious, but we still don't know how such a brain leads to consciousness in any animal.

We also know from the slime mould, that some sophisticated mind-like behaviours do not always require the neurological hardware of a brain. The slime mould's observed behaviour, its exploration of its surroundings, its ability to learn and to respond to its circumstances, is not so dissimilar to what we would expect from an animal with a skull packed with millions, or even billions of neurons.

Perhaps there is some base awareness found in the double-slit experiment at just the level needed to deal with the options presented: whether a photon should behave as a wave or as a particle; whether it is part of an observed system or not. If there were an element of mind everywhere, if the entire universe were mind as well as matter, inseparable and one and the same thing, then there would be no such thing as aware vs unaware matter. There would only be a difference between the circumstances that matter finds itself in, a reaction to which might then be observed as a form of behaviour. So it is to behaviour that we should look.

As I said in Chapter 1, however, it does not necessarily follow that inanimate objects like rocks should be said to *be conscious*. A rock could be full of molecules that react in the same way that photons, electrons and buckyballs do when presented with two slits.

However, that does not make the rock an aware entity, because the rock never acts to maintain its existence. It never acts as an independent entity. It has no rock behaviours.

This chapter is certainly not suggesting all quantum phenomena should be considered as evidence of mind in all things. There is a tendency for weird phenomena in physics, such as dark matter or quantum entanglement, to be given as explanations for the problem of consciousness simply because they too, are not fully understood.

For example, following Bell's theorem in the mid-1960s, the possibility of entanglement—where two particles separated by vast distances can have equal and opposite properties of spin—became an accepted fact in quantum physics. Some argue this means there is instantaneous communication between the particles, which they attribute to a universal mind. Perhaps experiment will one day show that the human brain is sometimes affected by matter or energy situated far outside of the brain due to quantum entanglement.

However, I am not convinced that quantum entanglement is the strongest example of how an element of mind would manifest in the quantum realm. For me, the quantum double-slit experiment seems a better fit for the question of mind in all things, because there is a behaviour to be observed, which looks a bit like decision making, born of awareness.

This also does not mean a mysterious force of mind needs to be added to the physicists' equations along with the known forces of energy and matter.

The idea of an element of mind in all things is not, in fact, an interpretation of quantum mechanics. It can only ever be a philosophical position, and I suspect is unlikely to lead to any new equations that make predictions. Instead it may help to frame what might be happening in a different context, perhaps give those making the calculations a slightly different perspective.

Wanting to view the world as aware because our species has disrupted the balance between humans and nature, does not alone make it a sound idea. In science, a hypothesis is tested by experiments, with the possible outcome that the hypothesis is false. In such a way, a hypothesis needs to be falsifiable to be considered fact, and is only fact when the attempts to falsify have failed. Some will rightly ask, not just can mind in all things be proven, can it also be disproven?

If the question here is, can we conduct an experiment to prove (and disprove) the universe is aware and intelligent, that there is an element of mind in its basic fabric, then I suspect the answer will always be no.

However there are many influential scientific ideas that are unlikely ever to be proven. Take the multiverse. How could an infinite, or near infinite range of universes that go with every interaction of every particle ever be proven? Whatever experiment might be devised to prove or disprove its validity, the experimenter may only be ending up in the universe that suits the outcome they are looking for. The multiverse is increasingly being considered a philosophical interpretation of strange phenomena, not a provable hypothesis. I should say, I am certainly not claiming the multiverse interpretation is wrong. I am only pointing out that it is really no more falsifiable than there being some aspect of mind in all things.

Some will think that making a distinction between an *aliveness* universe, with some mind in all things, and the observer effect, is splitting hairs, because both argue that awareness influences the experiment.

As mentioned previously, I believe the observer effect is really just a form of idealism. Idealism puts matter in service to mind, making it more acceptable to us than the notion of an intelligent or aware universe, because idealism partly re-enforces our species' bias that all understanding is centred on the human mind.

Picture once again the mind/body metaphysical scale as a continuum. Off to one side is materialism and physicalism, where matter comes first, and mind is the rare product of fortuitous conditions here on Earth, that have enabled the evolution of sophisticated animal brains. At the opposite end of the scale is idealism, where mind is the start and end point for all things. In the centre is dualism, seeking to unite two extremes. But also in the centre is the view that mind and body are one and the same.

Whether or not observations in this chapter on the double-slit experiment seem worthwhile, surely if some scientists are ready to go directly from the proven benefits of materialism to idealism, might this other, non-dualist, option in the middle also be worthy of consideration?

5

The non-problems of panpsychism

In 2008 the neuroscientist Anil Seth added a post to his blog with the somewhat indignant title, "Conscious spoons, really? Pushing back against panpsychism".

Seth was responding to an article where it was suggested some scientists were looking to panpsychism as a philosophy that could benefit those working at its margins. The article quoted philosopher David Chalmers as saying some forms of panpsychism would allow inanimate objects, including spoons, to have a form of consciousness.

The idea that being made of the same basic elements as inanimate objects, and our having consciousness, means we can assume some consciousness in all things, is in fact a minority view among modern panpsychists.

Most would see no evidence of, or purpose for, inanimate objects being conscious or even aware. I believe that if panpsychists argue all inanimate objects are potentially conscious, the philosophy will forever find itself relegated to the sidelines, as an eccentric faith-based idea.

I have said little about the philosophy of panpsychism itself and avoided using the term, where possible, as it has for many, connotations of New Age ideas that are neither credible nor useful. However, in this chapter I will briefly place it in the history of Western thought, and address the main criticisms any form of panpsychism must face.

As mentioned in Chapter 1, the simplest definition of panpsychism is the notion that there is some element of mind in all things. Anyone making a vaguely panpsychic case must of course elaborate on, and qualify what an element of mind in all things means in practice. This could be anything from full-blown subjective conscious experience, to the ability to respond to stimuli. It could perhaps, even include an entity having some experience without any evidence of a response.

I introduced panpsychism in Chapter 1, as being a highly flexible philosophical idea, that can encompass many related ways of thinking. It can be said to have aspects of animism and idealism, and even (unfortunately) of vitalism.

However, for me, notions of conscious spoons, tables or rocks are not really panpsychism. These are more a form of animism, which puts mind and spirit into inanimate objects as well as living organisms. I believe this is where panpsychism gets itself into hot water, attributing mind to entities that cannot exhibit behaviours. To the more sceptical, this looks like panpsychists are projecting mind into matter, without any evidence to back up their claim.

The majority of modern philosophies are monist, meaning they believe all things can ultimately be reduced to one. Scientific materialism and physicalism, for example, argue all can be reduced to matter. Idealism that all reduces to mind, and panpsychism that mind and matter are one and the same. Some would say even modern dualist philosophers are ultimately monist, because while some forms of dualism hold that there really are separate mental and physical realms, others hold that the mental and physical realms should be understood as different properties of the same physical reality, meaning there is one reality underlying both.

Some panpsychists have made their case from the perspective of *pansensism*, or *panexperientialism* which argues all things are on some level capable of sensing, experience or perception. Religious mystics believe God is everywhere, in everything, and that there is no separation between God and nature, an idea known as *pantheism.*

Despite these different emphases, pansensism, panexperientialism and pantheism all fall under the umbrella term of panpsychism, and are monist because the many things of our experience can be reduced down to one—although they differ in what they believe this irreducible thing is. In pantheism it's God. In pansensism it's sensing, etc.

That panpsychism is a monist philosophy is I believe its most important feature, and panpsychist arguments are more prone to failure when this is forgotten. For example, some view panpsychism as little better than the 18th and 19th century quest for an essence found only in living things, known as vitalism. Vitalism had many adherents before good solid science revealed the basis of organic chemistry. But vitalism was not a monist idea, because it assumed that a mysterious something *extra* was to be found only in living organisms. This divided the world into that which did, and that which did not, possess this mysterious essence. When panpsychism is not strictly monist, losing sight of the principle that mind and body (mind, matter and energy) are the same thing, it creates space for such accusations.

The roots of Western thought are usually traced back to Ancient Greek civilisation, which helped move Europeans away from anticipating the will of the gods, or placating nature with sacrifices, to a more objective understanding of the material world. For the Greeks, reason, deduction and observation were as important as belief and tradition, and as mentioned earlier with Aristotle's studies of the natural world, the job description for philosopher and scientist was really one and the same.

Arguably, throughout its 600 year history, panpsychism, as a refinement of animism, was integrated into early Greek thought, without Greek philosophers necessarily having adopted panpsychism in full.[39]

The Ancient Greeks commonly sought to establish what essential substances might compose all things. For Anaximenes it was breath, which allowed for the existence of *psyche*, the soul in all living things. For Thales it was water. Thales further believed all things contained different levels of psyche. He believed the lodestone, which can move objects by its magnetism, had more psyche than other entities, and that humans had the most psyche of all.

Empedocles believed all was made of fire, earth, air and water, and that behind these were the driving forces of *love*, which unifies, and *strife* which drives things apart. Empedocles theorised that at the beginning of the universe everything was concentrated in a unified whole of love, before strife broke this up and gave birth to the world we see around us, an idea with intriguing echoes of the Big Bang. Democritus believed everything in the world was made of tiny indivisible atoms, and that iron and water are different because they are made of atoms with different properties. Apart from atoms and their motion, all else was the void to Democritus.

Armed with little more than deduction there were some remarkable hits in Ancient Greek thought, like Empedocles' metaphorical Big Bang, and Democritus' early form of atomic theory. On the other hand, Empedocles' four essences—his concept of fire, earth, air and water—today sounds more likely to describe a heavy metal band's corny pre-show than the foundations of reality. Empedocles thought fire was the most powerful of the four, and perhaps seeing a blacksmith's forge transform metal into hard-wearing tools and weapons, for people with limited technology to examine their physical surroundings, it is understandable fire might gain the status as the most essential and transformative force.

Greek philosophers were variously monist in believing there was one essential thing, like breath or water, at the basis of all things, or more pluralistic, as with Empedocles, believing many things were at the root of reality.

The most enduring form of pluralism in Western thought is of course dualism, and for this reason the philosophy of Plato was well suited to later European thinking. In his theory of forms, he reasoned that because the physical world is imperfect and constantly changing, what we experience is really only an impression, like shadows on the wall of a cave. Plato argued that behind these shadows are perfect, abstract forms that have an existence independent of our perceptions. This is not, however, a world view based on a *yin* and *yang* of equal or mutually dependent spheres. The abstract world was perfect and accessible only by the intellect, whereas the concrete, physical world experienced through our senses was a distinctly imperfect one. Aristotle disagreed with the theory of forms, and even Plato later qualified his own theory.

Plato had a significant influence on early Western thought, particularly early Christianity, and not only because his was a dualist voice from the classical world. The theory of forms expressed a division between idealised and Earthly realities, similar to the way the Christian church sought to divide everything into the realms of God and Man. It also placed the abstract on something of a pedestal, so by association, the only intellect (the human mind) believed to be capable of understanding the abstract could also be placed on a pedestal.

However, to back up the point about the nature of Greek philosophy, there is also a significant thread of panpsychism even in Plato's work, such as this quote from *Timaeus.*

> This world is indeed a living being endowed with a soul and intelligence... a single visible living entity containing all other living entities, which by their nature are all related.

Moving forward, in 523 AD, a Roman administrator, Boethius, was imprisoned on trumped up charges for a year prior to execution. Boethius dealt with his unjustified loss of freedom and status by writing the erudite prison memoir, *The Consolation of Philosophy*. In the memoir he personifies philosophy as a woman who appears in his cell, and through patient dialogue, shows him that worldly goods, status and fortune are ultimately meaningless. Its message that reason based on faith, allied with modesty and humility, is all that has true value in life, meant it became one of the most popular books of the Middle Ages.

Yet the book lacks any examination of the nature of reality itself, and barely touches on any metaphysical questions. As the authority of the Christian church grew, and science developed into its own discipline, Christian philosophy became less concerned with enquiring what the physical world was made of. Instead its purpose was to present a world view where everything belonged either in God's realm of the spiritual, or the Earthly realm of Man. The purpose of philosophy was largely to understand what belonged where in this schema. Pre-Enlightenment, any attempt to unify the whole was not just unwelcome, it was also potentially dangerous heresy.

In 1979 Pope John Paul II made St Francis of Assisi (b 1181) the patron saint of ecology. St Francis' reputation as a lover of nature means many now regard him as a sort of mystical eco-saint, with aspects of animism and panpsychism, who just happened to be born into the Christian tradition.

In fact, St Francis joined the Crusades to the Muslim world, and through his Franciscan order, emphasised obedience to the authority of the Catholic church. Later protestants, including Martin Luther, even ridiculed St Francis for his conformity, seeing him as an example of all that was wrong with the authority of Rome.[40] However, those thinkers who did not adequately emphasise the authority of the church, as St Francis had done, were on far more dangerous ground.

Tommaso Campanella (b 1568) wrote two books in which he argued for sensing and knowing being intrinsic to all animals and all of the physical world, including the mountains and the sky. This sensing and knowing came from God, he believed, which is a blend of pantheism and pansensism. But he also believed sensing and knowing were inbuilt rather than emergent properties, which might lead to the conclusion there was no separation between the spiritual and Earthly, potentially ruling out God's will. This shading into nature-based monism was problematic enough for the Inquisition to give Campanella 27 years in prison. However, Campanella fared significantly better than his predecessor Giordano Bruno (b 1548).

Bruno wrote about an ever-present "world soul" of the Earth, and all the things upon the Earth as animate and sharing in this world soul. He also had a very modern understanding of the universe, which he saw as infinite. He further realised that in an infinite universe neither the Earth nor the Sun would be at its centre. Instead, the centre would appear to be wherever you happened to be at the time. For such dangerously modern notions, the Inquisition burned him alive in 1600.

While, as noted in Chapter 2, medieval society was strangely animist, finding spirits good and bad in all things, learned men mostly kept to the division between the Earthly and spiritual realms. Given the continued power of the church at the start of the Enlightenment, when René Descartes (b 1596) claimed that reason separated humankind from all other animals, he judiciously equated reason with soul, a soul that had been guaranteed by God.

Whether in private he was a true believer is perhaps not important. Descartes was an early voice for the power of reason. Using reason as an alternative to following inherited traditions made him a favourite of the French Revolution in the following century. But as previously described, he regarded nature as an unknowing, unfeeling mechanism, in which human capacity for reason was one of the few certainties. As such, in Europe, Cartesian dualism began to replace the division of the Earthly and the divine, with a separation of humankind from nature. Dualism was in a sense following a Christian philosophical approach, even as the power of the Church began to wane.

Baruch Spinoza was out of the Inquisition's reach, but still managed to get himself expelled from Amsterdam's Jewish community in 1656. He never declared himself an atheist, despite believing there was no personal God who listened to our pleas for help, that there was no afterlife, and the closest thing to God was in all of nature. This is surely the world-view of someone who would today opt for "Spiritual, but not religious" on his dating profile.

In the following centuries there were other notable philosophers with views either explicitly or implicitly panpsychic; Leibniz, La Mettrie, Diderot, Fechner, then in the 20th century Charles Strong and Peter Ouspensky. Even inventor Thomas Edison once wrote about "intelligent atoms". But as scientific materialism kept delivering technological breakthroughs throughout the 20th century, any outbreaks of panpsychic thought were usually overshadowed by the unrelenting march of materialism, which tended to increase our philosophical separation from nature.

The brief overview of panpsychism given above is only panpsychism as found in Western culture. There are ancient, and still active philosophies and spiritual traditions all over the world, which have had some influence on Western culture, and today elements of panpsychism are far more recognisable in Eastern than in Western thought. I have not attempted to give an overview of panpsychism throughout the world's philosophical traditions, as this book is about why this philosophy can be problematic for Western thinking specifically—and it is Western thought which currently has technological and economic dominance.

Just as Christianity has greatly influenced how people in the Western world live and think, India, the Far East, Africa and the Islamic world have retained many aspects of philosophy from their spiritual and philosophical traditions. A comprehensive overview of thought across the world can be found in Julian Baggini's book, *How The World Thinks*. What Baggini finds as the most significant difference between Western and Eastern thought is that, in the East, there is an assumption that language, logic and reason can only lead to a certain level of understanding.[41] In the West, some believe language and logic are a valid way to establish fact. In the East, language and reason are tools for reaching a point of understanding, rather than affording knowledge per se.

Although the majority of the world now measures progress by how fully a country has embraced Western economics and technology, Chinese and Japanese society in particular retain strong elements of their traditional ways of thinking. To this day, this includes aspects of panpsychism and even animism. Baggini points out that in the East, human life is more likely to be perceived in context with nature, rather than nature being the background to our existence. For example, one study showed that when viewing landscape pictures, Westerners tend to focus on the foreground figures, whereas viewers in the Far East are often more interested in the background and the space between objects than the objects themselves. In Japan, the love of lying on the ground and gazing up at the spring blossom is not simply a case of revelling in the visual feast of the cherry blossom. It includes the knowledge that the blossom will only last a couple of weeks, and will disappear almost as rapidly as it appeared. Cherry blossom is therefore an example of transience and emptiness, a reminder of the ever-changing, imperfect real world. For such reasons, there being some mind in all things, and an aliveness to the whole universe, is a less bizarre concept in Eastern than Western culture.

Except when debating ethics, later 20th century philosophy largely followed the materialist lead by concentrating on the human brain, its capabilities and whether any wider metaphysical conclusions can be drawn from advances in psychology and neuroscience. Based on neuroscience, however, some limitations on the ability of science to answer some of humankind's more metaphysical questions, have become apparent.

In 1974, the philosopher Thomas Nagel famously asked *What is it Like to be a Bat?* Because human and bat brains differ, and the bat can fly and map out its surroundings with echolocation, its lived experience will be significantly different to ours. Nagel's point was that whatever the essence of being a bat might be, it is probably quite different to the essence of being a person. Further, the bat's experience will likely remain unavailable to us, no matter how well science documents the bat's brain and sensory systems. Other famous philosophical arguments address this same issue, notably the *hard problem of consciousness* (which I will come to). First, let's consider Mary's room.

Frank Jackson asks us to imagine a brilliant neuroscientist named Mary, who knows everything there is to know about the brain and how it senses colour. However, Mary has lived all her life in a monochrome room, so has never actually experienced any colours for herself. One day she is liberated from her colourless world and sees colour for the first time. The question is, would her experience of, say, the colour red, be new knowledge? Would it differ from the factual information she has collected about how the brain experiences the colour red? In which case, facts about the brain are simply that, fascinating and illuminating information, which cannot fully replace our first-person subjective experience of the world. So the materialist and physicalist confidence that we will understand human conscious experience when we finally have a full and complete set of objective facts about the human brain, looks misplaced.

Mary's monochrome life has provoked plenty of debate. Firstly, there are technical issues with this thought experiment, such as the near impossibility of someone experiencing life solely in monochrome. This can be addressed by assuming Mary has a neurological disorder that allows her to only see in monochrome. Then her liberation is not leaving a colourless room, but surgery that allows full colour vision for the first time.

Another point that needs clarification was addressed by materialist philosopher Daniel Dennett who, in *Consciousness Explained*, notes that Jackson's wording states Mary has all the *physical facts* about colour and the brain. This differs from how it is often interpreted—the best available knowledge at the time—which will depend upon current technology. On the basis of Mary having all the physical information, Dennett concludes she would not learn anything new from her experience.

There are other ambiguities in this thought experiment. For example, in this context what do terms like knowledge and physical facts mean, and what does it mean to *learn something new* anyway?

I have no intention here to examine the for-and-against arguments around Mary's room, which could easily fill a book, let alone a chapter. However, it should be noted that Jackson himself believes a type of physicalism does, in fact, answer the question. Those of us who are not committed physicalists or materialists would I suspect agree with the view put forward by others, that Mary's first experience of colour is a different *type* of knowledge from anything she could acquire through study.[42]

What is important about both Nagel's and Jackson's thought experiments is how they lead us to consider the potential gap between the externally available information (the objective) and our personal experience (the subjective).

The other well known example of this is David Chalmers' *hard problem of consciousness*. Chalmers believes that in studying the mind there are relatively easy problems to solve, such as tracing visual activity through the nervous system to regions of the brain where visual processing takes place, or how the sense of smell or touch operates. But these may not answer the wider metaphysical issue, the so called hard problem of how sensations of the outside world, sensations of our own body, our memories and experience, all become our subjective experience. How do we go from a brain that has brain activity and neurons that react to the colour red, to having a subjective experience of red, and a sense of the redness of the red?

The hard problem is in part a re-framing of Nagel's 1974 question about whether more facts about the brain will ever answer the mind/body problem. Several scientists have entered the debate about the hard problem since Chalmers first presented it in 1995. However my impression is that most materialists see the hard problem as a distraction, and those who are more critical regard it as philosophical navel gazing.

The hard problem is a useful way of expressing the mind/body problem. However, taking it as more than that, could in my view lead to a modern-day version of Plato's theory of forms. It could be saying that the redness of the red —the abstract first-person experience—is harder to understand and therefore more of a mystery than the brain processes involved in seeing red. I would suggest that the important account we lack is not how red becomes red*ness*, but the more basic question of how conscious beings have emerged from apparently unaware matter.

The importance you attach to the hard problem does, I believe, depend mainly on whether you are inclined to a monist or dualist perspective. Rightly or wrongly, I think the hard problem more or less disappears for anyone with a strictly monist point of view.

Idealism, materialism and panpsychism are monist philosophies. For the idealist, the mind is ultimately all there ever is, so redness arises from the inventiveness of the mind. Physicalists and materialists believe first-person experience is only a change of brain state, which will eventually be accounted for by neuroscience. The monism of panpsychism means that the mind and body are understood as the same thing. But because we are considering the importance of red*ness*, of the subjective first-person experience and how it relates to the objective external world, I want to use the next section to make an observation about the constructed nature of subjective experience, by looking at one of the key moments in the child's development towards adult abstraction.

As children grow they reach certain developmental milestones as they gain full motor control, and begin to test the limits of influence over their parents (the terrible twos)! They start producing the language they have been soaking up since birth, and around this time will start to recognise themselves in a mirror. After a few more years, children become aware of their own mortality and perhaps start questioning the existence of a deity, and the possibility of an afterlife.

Perhaps in your childhood, you too had a moment like this clear memory I have from the age of eight? I was standing in my older brother's room looking at his model railway. Without warning, my developing self-awareness brought about this sharp and disturbing question: "If I am looking at this, who or what is inside my head witnessing that image?!" For a moment I had the notion of the world being watched by another someone (the *real me?*) sitting in a sort of mental movie theatre behind my eyes.

This notion of a little human, a *homunculus* sitting inside our heads is clearly wrong, although still slightly unsettling. We know it is wrong, because the inevitable follow-on question is then "if there is a witness in my head, is there then another witness inside their head...?" Because this would lead to a never-ending chain of subjects and objects, almost as quickly as the question is asked, the developing mind abandons this line of enquiry. It is instinctively recognised to be an unhelpful overuse of our mental capacity to abstract our world by separating subjects from objects.

Having asked this question, and instinctively realised this chain of subjects and objects is a dead-end, we return to what appears to be the most pragmatic way to understand the world. We treat ourselves as a singular, semi-permanent subject *in here*, looking at the world *out there*. This means the homunculus hasn't entirely been ruled out, and this idea that a real me lives somewhere inside my head is what Daniel Dennett calls the "Cartesian theatre".

I believe the experience I've described contains a wider lesson that is often missed. It should remind us that even this basic level of subject/object perception—me, the subject, separated from the external world, the object—is a selective picture of reality. It is a point of view we learn to adopt, a product of the way our hunter-gatherer brains operate, which allows us to experience the world. We can then reflect on that experience, to help the hunter-gatherer body navigate its way through life.

Returning to the hard problem, I believe that this question about subjective experience—how does the mind experience the sensation of redness—can easily become a search (in a metaphysical as well as neurological sense) for the core of our being. It becomes a search for the place where sensations of the external world finally *end up*, i.e. where do we experience the world from? In a way, this risks looking for a non-existent *real me*—the little man in the cinema.

To me, a far more significant question is how can we bridge the gap between a universe stuffed full of unknowing matter, to complex human experience, when rocks, mountains and people are all made of the same stuff? I believe the solution to the mind/body problem is to stop looking primarily inward at the human mind, and look outwards at everything that surrounds us. We need to give proper consideration to the behaviours and awareness in everything that is *not* part of the human brain.

I said the hard problem is not really a problem for monist philosophies, including panpsychism. You may have come across a phrase along the lines of *you are the universe experiencing itself*, sometimes attributed to writer Alan Watts, who immersed himself in Buddhist and Hindu philosophy, as interest in Eastern thought grew among Europeans and Americans in the 1950s and 1960s.

This expresses a common thread in Eastern philosophy and religion, about the inherent contradiction of existence. If you have been brought up in a Western mode of thought, you may wholly or partly agree with the principle of a universe experiencing itself, yet feel that although it contains an element of instinctive truth, it is ultimately too esoteric and enigmatic a concept to fit with rational philosophy or scientific evidence. Others will dismiss such thoughts as phoney New Age thinking.

However, I have not used these phrases here in the hope that they will trigger some moment of revelation about the nature of things, like a Zen koan that leads to great insight by suspending the rational mind. Instead, they put in context how we might use language to describe what an element of mind in all things could mean, in practice. If intelligent awareness is everywhere to some degree, and no more or less essential than matter and energy, then the first-person experience is only one impermanent point of view inside that wider awareness.

A phrase like "the universe experiencing itself" is, of course, not without ambiguity and possible contradiction. For instance, it might suggest all this first-person experience is being sucked into some central point in the universe somewhere, because our own experience is that consciousness comes with a singular point of view, located inside our heads. It might also hint at there being a purpose to existence, that perhaps there is some kind of divine intent that gives a *why* to human existence. As I said in Chapter 1, I think we should address the *how* of existence, rather than the *why*.

Some imagine a universe experiencing itself would mean that the individual consciousness therefore taps into a limitless store of universal knowledge, some sort of universal consciousness. Many panpsychists claim a universal consciousness must exist, which perhaps comes into being when people take part in a group activity, sitting in a lecture theatre, or singing in a choir, for example. For me, these are speculations that ignore the known limitations of the human brain, and suggest multiple consciousnesses are being linked by some mystical force, for which we have no evidence.

Although a philosophical view that an element of mind is everywhere may increase the theoretical possibility of there being a universal consciousness—and the same here applies to telepathy, remote seeing, and other paranormal phenomena—I do not believe there is evidence that minds can come together to form a distinct, unified consciousness, at least not in a way that could not otherwise be explained by sociology or group psychology. Whether or not evidence for any of these comes in time, evidence for human minds being somehow linked together telepathically is not something a view that there is some mind in all things, is dependent upon.

Again, for me panpsychism is essentially a monist philosophy, which means the many things can be reduced to one. Many things plainly exist. There are 118 different elements in the periodic table, 7.9 billion people on the planet. Materialism and physicalism are monist in saying that everything is ultimately physical matter. At the same time, the purpose of science is to more precisely differentiate the variety of things that make up that material world from one another. This means we must accept a degree of contradiction when we reach the limits of language, and this is the only route to making a workable panpsychism. Accepting contradiction in language, and our ability to conceive the basic nature of things, is key to making panpsychism comprehensible.

There is a basic contradiction in the notion of a universe *experiencing itself*. An entity that is experiencing itself, a mouse, a human or whatever, is experiencing itself as different from everything that surrounds it. Obviously I am a different entity from my surroundings, while at the same time my form and my survival is completely dependent on those surroundings. Therefore my sense of separation is serving a practical purpose, rather than describing an absolute reality. Cosmologists are uncertain whether the universe is finite or infinite, but the word "universe" is usually taken as referring to the entirety of matter and space. But if the universe is the entirety of everything, how can the universe separate itself from anything? There would be nothing outside of the universe to be separate from.

That panpsychism is a fully monist philosophy is also key to addressing what is supposedly its greatest challenge, the "combination problem". The combination problem asks, if the universe has an element of mind everywhere, how could these small units of mind combine to form a larger unit of mind, like human consciousness?

To me this is really a misunderstanding of an element of mind being in all things, but is worth addressing, as even panpsychism's advocates sometimes regard the combination problem as a significant barrier.

Imagine a sunny day at the beach, and you have arrived with a bag full of volleyballs, ready for a tournament. It makes no difference if the volleyballs are inside a net bag, on a table, or lying loose on the sand. Even if you were to painstakingly glue them all together, your separate volleyballs will remain separate volleyballs, and not morph into one large shiny new beach ball. To achieve that, you would need to take them to a recycling facility, extract the plastics and re-manufacture the raw materials into the shape of a beach ball.

But if we look up from the beach to the waves coming into shore, a wave is not a fixed thing with boundaries and limits like a volleyball. Does anyone believe a large wave is composed of many smaller distinct waves? While waves large and small have similar properties, they are only unique arrangements of matter and energy at a specific point in time. My point here is, if mind and consciousness were like volleyballs the combination problem would certainly exist. However, the sense that each human or animal consciousness is a separate and distinct entity is only the result of our brains imposing order on all the unordered stuff around us.

So what do we mean by mind? Unlike a brain, there is no single object called "the mind", with boundaries and limits that can be identified. Internally, mind refers to our first-person experience of sensory information and thought processes, which may or may not produce actions. Externally, mind is indicated by those behaviours and responses we can attribute to a mind. For example, my lower leg jerks when the doctor taps me below the knee, and as a reflex reaction, we say the mind is not involved. Whereas if I then raise my lower leg because I want to show her a bunion on my foot that I'm worried about, the conscious mind is involved. The reason we tend to think of mind as one thing, more like a volleyball than a wave, is because we live almost constantly through the homunculus-like, singular point of view, separating ourselves off from our surroundings.

If awareness truly infuses the base nature of the universe along with matter and energy, then there is no indivisible smallest unit of sentience, any more than there is a smallest unit of matter or energy.[43] In fact, I would argue that any form of panpsychism which recognises distinct and separate units of consciousness at the smallest level is not really panpsychism at all. It is really a type of dualism, because the properties of mind are then fundamentally different from those of matter, which we know can be combined easily into different forms.

If we assume a mind is a discrete entity produced by a single brain, at some point it would cease to be a mind if we removed enough of its components. Take out language, meta-knowledge (knowing what you know), capacity for feeling physical pain, for experiencing joy, sadness or hope, and the label may no longer be justified. I suggest this is the error that leads to the combination problem. It is that we start from the perspective of the human mind, then trying to divide this down into smaller and smaller units. Smaller and smaller human-like minds become progressively harder to conceive of, as we work down the scale. Instead we should look at our consciousness as a temporary point of view built from the ground *up*, from the awareness in the subatomic and atomic level, then the organic chemistry, etc. All of which gives us reason to reject a combination problem in panpsychism.

Neurological evidence also suggests that mammalian brains do not always produce a single mind. In 1981 the neuroscientist Roger Sperry won the Nobel prize for his work on split-brain function. In binocular mammals, half the signals from each eye are sent to each of the two hemispheres. In experiments with monkeys and cats where the two hemispheres had been separated by severing the corpus callosum, Sperry found the animals' hemispheres often functioned independently.

With cats he covered one eye with a patch and taught them to reach for a food reward when they saw a particular pattern (either a cross or a circle) on a wooden block. While the cats were learning to associate a specific pattern with food, the eye patch ensured they only saw the block with one hemisphere. When he later covered the other eye, the cats did not reach for food when they saw the familiar pattern. This showed that new information learned by one hemisphere was not available to the other.

The monkeys and cats were of course not volunteers, unlike Sperry's human subjects, whose hemispheres had been separated as a treatment for severe epilepsy. One of the most striking results came when he made a word available to one side of their visual field, and an image to the other. He then asked the volunteers to both name and draw the object. They correctly identified the word, and then drew a copy of the image they had been shown. However, because the two hemispheres were operating separately, they were unaware that the word and the image represented two different objects. These patients lacked a unified experience of the word and the image. Later research has found a similar split-brain experience is possible with the auditory sense.

The split-brain hypothesis is sometimes simplified into a popular myth that the left hemisphere is logical and language based, whereas the right is emotional and creative. Memes circulating on the web often claim that the way we view an optical illusion can diagnose which of these two dominates our character.

The reality, of course, is more subtle and complex than each side having strictly delineated functions. In cases where one hemisphere is damaged, functions may partly be shifted from one hemisphere to the other. The split brain can also be understood as a war of the hemispheres, which compete to dominate our attention. That said, they can have quite distinct characters, which is something I will revisit in Chapter 8 with the story of Jill Bolte-Taylor.

We cannot know for sure whether people whose hemispheres have been disconnected are in some way living with different experiences at once, or perhaps switching between two points of view.

To understand another's experience usually requires some account of their experience, given through language, which for people who have had this surgery may only represent the world of one hemisphere. However, whatever the exact neurological function of each hemisphere, one thing we do know is that the two different hemispheres can take significantly different perspectives inside the same brain.

The main characteristic of true consciousness is perhaps that it is the final moment where different sources of sensing and information are brought together as a unified experience. The fact that the hemispheres can function independently when the communication between them has been cut, leads one to question whether conscious experience is actually a fixed, unified or indivisible thing.

Returning to the combination problem, it is among philosophers still considered a live issue for panpsychism. But the combination problem is partly a legacy of centuries of panpsychism in religion and then science, where the soul, mind and consciousness were sometimes considered interchangeable.

I personally have never believed in an immortal soul or re-incarnation, or believed there is some essence of me that makes its way from life to life, or from this life to an afterlife. Consequently, I have no difficulty in viewing my physical body and my consciousness as equally temporary and transient forms that will lose their identity with the death of my physical body.

Just as my body is not fixed, absolute or unchanging throughout my life, neither is the first-person experience that I identify as *my* consciousness. This has always made the combination problem something of a mystery to me. To repeat the main point, if mind, matter and energy are all the same basic stuff, and the physical world is easily combined into different forms, then why would that not also be the case with mind? The combination problem occurs when we treat the mental and physical worlds as two different things, which is a dualist answer, not a monist one.

Another answer to the combination problem came from American philosopher and psychologist Charles Strong. In his 1918 book, *The Origin of Consciousness* he considered the cells and structures of the human body as potentially full of small moments of experience, yet argues the mind is unable to perceive and deal with all these interactions, so instead experiences a singular point of view. That may sound speculative given he was writing over a century ago. On the other hand, our physical senses and our perceptions of the physical world are too approximate to perceive solid objects as collections of trillions of individual atoms. Solid matter is, in fact, over 99% empty space, something we are not capable of perceiving.

The perceived solidity of the physical world is correct in a pragmatic sense because the human body cannot pass through solid objects, and our ancestors knew that if they smashed a nut with a rock the impact would break it open and give them food. Our perception of the world, and the experience we gain from being in possession of a mind, does not represent absolute reality. It is just what is useful, what is good enough for the purposes of the hunter-gatherer body.

In 1974, the same year as Thomas Nagel's essay, another significant philosophical essay for understanding human consciousness was published: "All animals are equal" by the Australian philosopher Peter Singer. Here, Singer questioned the rational basis for the human race's frequently appalling treatment of other species. Singer is himself a vegan, humanist and long-time animal rights advocate. His essay addressed the many logical flaws in our justifications for exploiting other species. For example, the idea that we have the right to use animals because we are more intelligent. Singer points out that we don't assign rights by intelligence level within our own species, and an adult pig is considerably more intelligent than an infant human, but only one of these ends up on a menu.

Singer concludes that a view on the rights of animals must take into account their potential to suffer, and that ultimately our current treatment of the rest of nature shows a large degree of species-ism.

However, Singer's essay has for me an additional significance beyond being a thought-provoking work of moral philosophy. It has a role in the wider metaphysical debates about human consciousness, because if on analysis our justification for placing ourselves above the other species is inherited and cultural, rather than based in fact or reason, surely it is also wrong to regard human conscious experience as so removed from the rest of the animal kingdom, and from nature?

The myth of human exceptionalism is surely incompatible with panpsychism. While recognising the incredible capacities of the human brain, I have been arguing we should also consider human consciousness, and our highly subjective first-person experience, as a subset of awareness in an ultimately aware universe. If mind is in everything, as panpsychism argues, taking a ground-up approach means mind is ever-present, routine, almost mundane. In fact, as we will see in the next chapter, our senses and perceptions are frequently flawed, and in many ways, human consciousness is an approximation of reality, not the most elevated form of awareness.

To give an alternative, ground-up way to regard the place of human consciousness, I'll briefly draw a parallel between human consciousness and a class of legacy software called a *screen-scraper*.

Imagine two computer systems exchanging data with one another. System A is trying to create a new customer record, with first name, last name, title, customer email, etc. in a customer database on system B.

Computers usually communicate through agreed standards, using web services (Rest or SOAP) or an Application Programming Interface (API). Once system A has connected to system B and authenticated itself, one simple message, correctly formatted, will create a new customer record in the database on a server that could be thousands of miles away.

However, what if your organisation still possesses a legacy system from an earlier age of computing? Before affordable scanning and text recognition, many older systems were designed on the assumption that a person would be sitting at a computer terminal, inputting data from paper forms. As a result, many legacy systems lack the interfaces needed for direct machine-to-machine interaction.

Rather than going to the expense of reprogramming legacy systems, in the early 2000s it was not uncommon to use a type of program called a screen-scraper. Screen-scrapers read the screen output that was originally intended for human consumption. Like a person, the program navigates through menus, reads text, puts data into input boxes and presses buttons, behaving as if it were an operator sitting at a terminal. It is of course inefficient and error prone, as the program must constantly wait for the legacy system to deliver the screen output, then check that the expected elements are where they should be on the page. It must have subroutines ready in case something unusual happens that breaks the expected workflow.

In this sense, a virtual world—a sort of virtual operator—has been created between the two systems in order for them to interact. The screen scraper is acting as a proxy for the absent operator. I see this as a metaphor for the highly subjective and imperfect first-person consciousness in our heads. Our minds create a sense of self, a virtual operator, that sits between the physical world and physical body.

Human exceptionalism regards human consciousness as the brightest light of all animal awareness. However, our brains are also producing a frequently flawed first-person experience. Similarly, the screen-scraper mediates, slows down and limits how two systems interact. Perhaps out of necessity, our rough-and-ready human consciousness is also mediating, limiting and slowing down our interactions with the universe? The complexity of the human brain, with its billions of interconnected neurons, might therefore be seen more as a *mediation machine*, rather than being held-up as nature's greatest achievement.

We have encountered a life-form with a less mediated interaction with its world, namely, the slime mould. We know it can learn, anticipate future events, and navigate mazes more successfully than many basic robots. It can choose between 11 foodstuffs with a varying balance of carbohydrates and proteins, and pick the one best suited to its physiology. It can map out transport networks using zero neurons, compared to the billions available to us for the same task. Now, this does not mean the slime mould *is* having a first-person experience as it makes dietary choices, or maps out a network. The slime mould appears to love oats, but we cannot know whether it experiences any *oatiness of the oat flake*. This is unlikely to be established with any confidence, because there are no parallels to be drawn between the slime mould's internal processing, and the processing of the mammalian brain.

Yet the *oatiness of the oat flake* is not a wholly whimsical idea, as oats are not all the same to slime mould. One group of Australian researchers found the American slime mould they had been sent was uninterested in the locally sourced organic oat flakes they provided, but tucked-in once it was given a well known corporate brand of American oats!

We don't know why that was the case. But we do know slime mould exhibits behaviours that rely on skills such as learning and memory, that until recently were only credited to animals like us, with sufficiently complex brains. Slime mould being a picky eater with brand loyalty, having a preference, means we cannot rule out potential experience. Of course it does mean there *is* experience. But from the outside its behaviour suggests a liking, and in humans we would say a liking goes with having that experiential sense, the something *ness* of a preference.

Some panpsychic philosophers would go further, and say first-person experience must be happening down at the level of basic matter, inside the slime mould and its atoms, and they believe this is what every panpsychic should be arguing for. Personally, I see this as problematic, as it makes a broad assumption that universal experience is a logical necessity, rather than being backed up by any behavioural evidence, which at least can be assessed and debated.

Panpsychism requires a level of humility about the significance of human consciousness. Maybe human consciousness is a cumbersome, mediated way to interact with the world? The slime mould could be said to interact more directly with its environment, making it more like the efficient web server, than the inefficient screen scraper. In terms of processing power relative to behavioural complexity, lacking even a single neuron, the slime mould could even be judged as far ahead of us.

A final philosophical problem for panpsychism is whether it has implications for artificial intelligence. For example, would this philosophical view make self-aware robots more or less likely? Does it, somehow, make a case either way for whether AI should have rights?

I said in Chapter 1 that this book would not address AI in any detail. This is because the question at hand is the mind/body problem of human existence, and how animal consciousness like ours can be a feature of an apparently unknowing universe. I am observing the widespread extent of intelligent and aware behaviours in nature, not attempting to describe the nature of human consciousness itself. Trying to describe consciousness in a way that accounts for all the possible permutations of AI, now and from here on in, would make the mind/body question virtually impossible to answer.

Some will think a book that does not extensively debate the implications of AI is not properly dealing with consciousness. To me this is rather like saying "Okay, so that's a great description of how snow drifts form in the Arctic. But it should also describe how snowdrifts would form on a far distant planet with a completely different chemistry, gravity and atmosphere, the parameters of which are currently unknown." They are two separate problems. For one of them, the basic facts of the world are historic and knowable. For the other, there are simply too many unknowns.

Still, one worthwhile point relevant to panpsychism does come out of the AI debate: that experience and comprehension do not necessarily depend on neural complexity.

While I am arguing that basic organic life, and matter itself, may be said to have some element of mind and awareness of its environment, I do not believe a silicon-chip based computer is aware of itself as an entity, any more than a rock is aware of itself as a rock. It is true that a robot could be programmed to behave like a living being. It could try to preserve its existence as organic life does, and may one day pass the Turing test and convince people that it, too, is a living being, despite having no inner life. These can be achieved without the machine having any sort of experience.

At the other end of the scale, if a synthetic brain were grown in a lab, there is a distinct possibility it *might* be aware or conscious, because growing a brain replicates a natural process, which is different from programming a computer. Where things get complicated are all the points in between, and the endless possibilities that will be created by future technologies.

Coming back to the human brain, some claim we are the only animal that should be credited with a really meaningful form of consciousness, because evolution has led to the human brain's unique size, structure and complexity. This means thanks to evolution, brains in the animal kingdom reached a threshold where they became sufficiently large, and their neurons sufficiently interconnected to enable self-reflection, forward planning, meta-knowledge (to know what I know) etc., which then count as true conscious experience. At the same time, many animals are able to perform apparently complex tasks without necessarily understanding them, which Daniel Dennett has neatly described with the phrase, *competence without comprehension.*

However, additional complexity in human technology does not necessarily lead to an increased likelihood of awareness. The computing power that got Apollo astronauts to the moon in 1969 is exceeded many times over by the most basic electronics in our homes, including the USB-C charger. Clearly there is a vast difference between the processing power of a USB-C charger, or toaster, or car key fob, and that of a modern supercomputer. Still, these all process data without any understanding of what they are dealing with. They are only mechanisms for the transformation of data from one state to another. They have oodles of competence without an ounce of comprehension! The supercomputer can perform multiple tasks far more rapidly than a key fob. Yet the supercomputer has no more understanding of the data it is transforming than the key fob does. Assuming basic silicon microchips are not aware of themselves, the extra complexity of the supercomputer's circuits brings no understanding or insight into the mix. Complexity does not cause comprehension in this case.

Then look at animal consciousness. We have an animal, *Homo sapiens*, we can confidently assert has a significant level of first-person experience. We are a species that has evolved from simple organisms, through early marine animals, then via a shrew-like creature to land mammals, primates, early humans, onto the modern human. The emergentist argument for keeping humans above the other animals is that somehow, thanks to evolution, either neurological complexity, or some special structures in the brain produce first-person experience and understanding in complex brains like ours.

But the successful production of consciousness by mechanically engineering a brain with human-level neural complexity would be inconsistent with what we find in computing. In computing, complexity changes the rate, difficulty and variety of tasks performed, yet there is still an absence of understanding inside the computer. From the panpsychic point of view, therefore, it is not wildly outrageous to think that if we are having a first-person experience, there could also be, at least a trace of, first-person experience in other animals, perhaps even in bacteria or the no-brained slime mould, and that awareness is fundamental to living entities.

The materialist response may well be: "Well, first-person experience is nothing special, just another program, which needs hardware of a certain level to run." We could respond that an 8-bit computer from the 1980s can produce video images with a very low resolution compared to the ultra high definition videos a smartphone can deliver. Yet both are only playing video, producing moving images that differ, as Darwin said of the mental worlds of different species, in extent rather than type.

The reply here might be: "Ah, but a 1980s computer could never run a 3D video program, because it would be limited by its hardware", and so on... Again, I don't see these AI arguments really resolving anything here. Ultimately, the side people choose will depend on whether or not they believe the existence of consciousness depends on a specific, complicated brain structure.

In the next chapter I will describe the brainstem theory of consciousness, which asserts that any animal with a brainstem is potentially a conscious being. The brainstem in humans is a far smaller and less complex structure than the human cortex. This is another reason to question whether the level of complexity in the human cortex is really the key to consciousness.

6

Where in the brain are we?

Except in extreme cases of mental illness, no matter how much the world changes around us, we appear to live our lives as a singular self. The same person whose narrative is suspended when we go to sleep at night resumes its story when we wake up in the morning. If we see a photograph of ourselves, whether it is a high resolution digital photo taken 30 seconds ago, or a grainy print from a disposable film camera taken 30 years ago, the same person is being represented. However much we may have changed through our experiences, our current behaviour and character appears to exist as an aspect of a singular self, one that has a story. Many believe the essence of themselves also has an existence outside of the physical.

We appear to be the final destination for sensations that stream into us from the external world. But that idea of sensory data streaming towards some core of the self is not the preferred model of many neuroscientists, who now describe human perception as a form of *controlled hallucination*.

I have suggested that there may be an element of mind at many levels in nature, and that we should consider human consciousness to be a subset of a more pervasive awareness. If awareness is all around, and we are only one point of experience within that, how would this relate to current neuroscientific views of consciousness?

First, I want to look briefly at a couple of points from neuroscience, which show how constructed this experiential self can be. I will then look at why some well-respected scientists now argue that consciousness is likely present in nearly all animals, and emerged in living beings as far back as half a billion years ago.

Because vision is our primary sense, it is a major area of study for psychologists and neuroscientists seeking to understand the relationship between our brains and our experiences, so I will look briefly at how the physical world in front of us becomes an image that we perceive, and how that image might become something of meaning.

Broadly speaking, the process of vision works like this: sensors at the back of the eyeball turn light into electrical signals, which travel via the optic nerves to the thalamus in the middle of the brain. The thalamus does some initial processing, before the signals continue to the left and right primary visual cortices at the back of the brain.

There are then some 20 different areas of the brain involved in visual processing. Neurons in these regions can have specialised functions, some being activated by the edges of objects, others by direction of movement. Some will react to colours, being activated by red but inhibited by blue, while other neurons will do the opposite.

In terms of transforming visual input into meaning, what is perhaps most notable is that the two streams of information exit the visual cortex with different purposes. Some head down and forward to the temporal lobe, where the brain distinguishes detail, and the features of objects, giving you the *what* of the world. The second stream goes from the visual cortex forward and up to the parietal lobe, which specialises in an object's location and its movement, allowing us to understand where things are in three-dimensional space. Then, through some process as yet to be determined, these two streams of visual information are linked with information from the other senses to create an apparently unified experience of ourselves in the world.

Naturally, that description is greatly simplified, and not only in terms of physical biology. The brain is doing far more than simply converting signals from the outside world into an experience. For instance, there is the blind spot caused by the lack of photoreceptors where the optic nerve and blood vessels exit the eyeball, which is always there but rarely noticed, because the brain constantly fills this gap in our visual field.

However, even accounting for the brain's ability to patch up blind spots, and incorporate vision with other sensory data, this account of vision is not only incomplete, it is in one vital aspect misleading—it suggests vision has a simple one-way traffic flow. It suggests there is a flow of light from the outside world into the eyeball, then onto the deep recesses of the brain, which then generates a sort of TV, or movie show in the private cinema of our experience.

Yet, in one important sense we are not even seeing what is in front of us at the present moment. Instead, because of the time required for visual processing, the brain must constantly predict what we *should* be seeing.

Chronostasis is, rather like déjà vu, a fleeting moment where we notice something out of the ordinary. Those of us (Generation X and older) used to seeing analogue clocks on the walls of doctors' surgeries, schools, offices and railway stations, may have, on occasion, looked at a clock and noticed the second hand apparently freeze, before the clock started ticking again. We may have experienced a moment where time was suspended, a moment that appeared unnaturally long, signifying that either the clock is faulty, or something peculiar is happening with our perception.

The apparent freeze happens because your brain knows that clocks are sometimes stopped. When you first see a clock, the brain cannot be sure whether this clock is running. It needs time to observe, because all that hunter-gatherer neurological hardware in our heads creates a significant time delay between the light entering our eyes, and the sensory input being turned into our experience of "the now". Until the second hand either moves, or does not move, for what your brain estimates to be a second, it does not know whether you should experience a moving, or a stopped clock.

This means, not only is our experience of the current moment subjective, strictly speaking it is not even the current moment at all. It is the brain's prediction of what the current moment will look like, a fraction of a second into the future.

Although we can celebrate our consciousness, praise it as *rich, wondrous* or *vivid*, chronostasis is also a reminder of how rough-and-ready it can be. Consciousness is pragmatic, and like most evolutionary endowments, visual perception is not about achieving the highest level of accuracy. It only needs to be good enough to allow the species to avoid extinction.

We are not passive receivers of sensory inputs, because both vision and hearing are active and participatory processes. I said the thalamus does some initial visual processing, but this is something of an understatement. There can be 10 times as many signals going from the visual cortex *to* the lateral geniculate bodies in the thalamus (where input from the optic nerves is first processed) as there are coming *from* the eyes back to the visual cortex.[44] The majority of traffic can in fact be going out from the visual cortex, away from the locations where we would expect experience to occur, not towards it.

Rather than the eyes being like TV cameras, and the brain a receiver, this traffic flow suggests the brain is rather creating its own TV programme, then splicing it together with another TV signal being fed in from the outside. It puts together these different sources, internal and external, to create our reality.

This internal generation of reality allows us to compare the exterior world to the world we expect to find, a form of error checking that takes place in the thalamus. An example of this is MIT professor Edward Adelson's checkerboard (pictured), which demonstrates how your brain uses the objective information coming from the eyes in order to produce perceptions that more closely align with its expectations.

Fig 6.

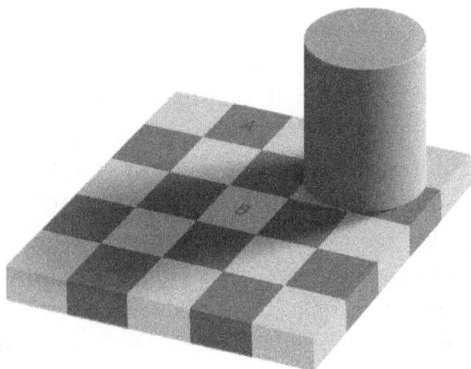

In both images (Figs 6 & 7), square A is the same shade of grey as square B. However the brain knows that chequerboards have alternating squares of light and dark. It also knows an object in shadow will reflect less light. Knowing the cylinder would cast a shadow, our brain makes square B appear lighter than it actually is. You can see they are the same in the second image, because squares A and B are connected by grey rectangles of the same tone.

Fig 7.

This is the sort of phenomenon that leads neuroscientists like Anil Seth, to liken our conscious experience to a sort of *controlled hallucination*. The world we experience is very much something it actively generates, created by our minds as much as by the outside world.

In 2005, it was discovered that your brain contains "grandmother" neurons, which fire when you recognise your grandmother. If you can recognise Halle Berry and Jennifer Aniston, your brain also contains at least one Halle Berry, and one Jennifer Aniston neuron. Not only will these fire when you see photographs of either actor in different contexts, they can handle more obscure representations. For example, in that same study, the Halle Berry neuron was activated even when subjects were showed a *cartoon* of Halle Berry as Cat Woman.

To us, faces are not the separated features of eyes, ears, noses and lips, cranium etc. We integrate all this information into a face, from which we read emotions and intentions. But how a neuron firing actually leads to the experience of knowing you are looking at Jennifer Aniston or Halle Berry is the still unanswered question of phenomenal consciousness.

In the mid-1800s, a responsible railway foreman named Phineas Gage experienced a profound personality change, becoming angry, impulsive and childish, after an explosion forced a metal rod through his left frontal lobe. Since then, this region of the brain has been understood as key to our self-control. One form of dementia, named frontotemporal dementia, affects the same region and can lead to a loss of the ability to plan, and a deterioration in language. More distressingly, it can also lead to significant personality changes, loss of empathy, impulsive behaviour, and increased aggression.

Given that degenerative disease and brain trauma are now well-documented causes of change in a person's character, for most of the 20th century the consensus was that the seat of consciousness would need to be located in the cortex. More specifically, it was believed to be in the frontal cortex, as that is the location of planning, reason and language, those supposedly unique abilities of the one animal we can confidently assert is conscious, the human animal.

It is probably useful here to identify two different parts of consciousness. There are the cognitive functions, such as language, reason, learning, planning, problem solving, etc. that give us the ability to process information and carry out tasks. But there is also the subjective inner experience, the feelings that come with being conscious, known as phenomenal consciousness, which are only available to the person experiencing them. In the broadest possible sense, these are the objective and subjective aspects of consciousness.

In academia, studies of the mind in the first half of the 20th century were heavily influenced by Sigmund Freud. What may surprise some is that Freud had practised as a doctor, neuroscientist and lecturer in neuropathology. In 1885, after witnessing some apparent breakthroughs with hypnotism, Freud came to believe that psychological techniques might be more productive than the existing rudimentary methods for examining the physical aspects of the brain. He apparently saw psychotherapy as a powerful tool, but also something of a stop-gap measure, until neuroscientific technology improved. Even so, he believed proper analysis of patients' subjective accounts could create consistent theories, and lead to an early science of the brain.

As neuroscience caught up in the mid-20th century, Freudianism, and along with it, subjectivity, fell out of favour. There was a desire for objectivity, and what was known as behaviourism came to dominate animal and human study. This was a very different approach to Freud's, because behaviourism was only interested in what the brain can do, not what a person experiences. It regarded lived experience as a black box of inputs and outputs, because subjective experiences could not be assessed or measured objectively.

From the 1960s, some psychologists and neuroscientists modified behaviourism into functionalism, which views the brain more as a set of algorithms, organised into functions like a computer. One supposed advantage was that this would allow for brain functions to be modelled with new technology, and so reveal what the brain was up to.

Strict behaviourism and functionalism went out of fashion, but they continued to influence attitudes to scientific study of the brain in the following decades. Although the desire to study the brain as an objective biological machine has undoubtedly led to great advances in areas like memory and learning, it was unable to account for, and largely uninterested in, a person's lived experience.

From the early 1990s, there was a move to address and overcome the perceived metaphysical gap in neuroscience, by directing attention to the subject of consciousness itself. Although neuroscience has continued to make progress identifying what brain activity correlates to a person's subjective experience (called the *neural correlates* of consciousness) it was around this time that David Chalmers formulated the "hard problem of consciousness". He believed neuroscience was only addressing relatively easy problems of the mechanics of perception, not than how these neural correlates came together to form our integrated subjective experience.

However, some disagree that there really is a hard problem. On the unanswered question of subjective experience, neuroscientists like Anil Seth make a parallel between this and the Victorians figuring out the chemistry that enables organic life. At the time, this also appeared to be a near impossible mission—until the correct science was done to figure it out.

As previously noted, until recently, when scientists and philosophers addressed consciousness, the majority of their attention was on the human brain and human experience.

There seems to be an implicit assumption that by completing our study of the human brain we will come to understand consciousness itself. Animal experiments are often conducted to investigate common mechanisms across species, in the belief that they would enlighten us more about our experience than theirs. Morgan's canon still held sway, and until the end of the 20th century, assigning experience to other species was largely considered problematic or unnecessary.

But in order to understand the concept of consciousness, which supposedly puts us at the top of the pyramid of intelligent evolution, we must account for more than just the human experience. If human consciousness really deserves a special status, we need to describe what it is we have that other species lack. This means having some idea of *their* lived experience, as well as their cognitive abilities.

Those philosophers and scientists who remain sceptical about the potential for true consciousness in other species often argue that the idea other animals can have meaningful inner lives is a sort of "folk wisdom".

Folk wisdom here is the idea that our hunter-gatherer brains have hard-wired us to project a theory of mind onto other species, in order to anticipate their behaviour for hunting and to avoid being their prey. The claim is, this then predisposes us to believe other animals must be having a conscious experience similar to our own, without any evidence that it is so.

Those who still believe human consciousness to be fundamentally different will argue that other animals are dominated by instinctual or learned behaviours that involve no real thought.

For them, this differs from true consciousness which they believe depends on attributes such as language, abstraction and an ability to plan. These allow the conscious being to separate itself from, and have control over, its more instinctual responses. I don't know whether this is the result of a strict application of Morgan's canon amongst some thinkers, or because they are suspicious of first-person experience in general, but they rarely attempt to *imagine* what the mental worlds of other animals might actually involve.

For instance, we have no evidence that crows have language. When the crow 007, described in Chapter 2, works his way through a series of different tasks that need to be completed in a specific order, and uses less trial and error than many humans would, does the probable lack of an internal dialogue mean this cannot be counted as abstract problem-solving?

Or does it count as abstraction, but a lesser form than human abstraction, because crows lack an internal commentary as they solve problems? We know the crow 007 has seen elements of the test before, so could he be visualising the different steps and changing their order mentally, in order to get the long stick needed for the last step? For those who say 007 is not thinking in the abstract, if that were so, what other process would allow him to link the steps required in order to perform this task?

As a species we know what it is like to be a human, through years of our own experience, which we can share by describing it to one another. Any objective, external description of another's consciousness can be related to our own experience through language. In this way, we always carry a kernel of understanding, thanks to our experience— something we carry regardless of any other theoretical knowledge we may have.

If we do not attempt some understanding of other species' experience, we give ourselves an unfair advantage. If we assess other species' consciousness by starting with human consciousness, then work to subtract abilities, it is unsurprising that other animal minds will barely seem to resemble our consciousness at all. Determining if other species are conscious, requires a notion of what the mental worlds of other animals could include.

While experimental evidence can give good reason for taking away some human traits from other species, we still need to have some idea of what it might *be like to be*, for another animal.

The dreaming brain

While we can use objective measures to understand other species' cognitive abilities, the other category of consciousness —the nature of their lived experience—will always entail a degree of inference and deduction. However the choice is not simply between rampant anthropomorphism on the one hand, and animal experience as an impenetrable black box on the other. Just as we can set tests that reveal cognitive abilities, we can use neuroscientific techniques to deduce the scope and nature of any first-person experience in the animal kingdom.

Dreams are a useful starting point, because they are made up of first-person experiences with little, if any, input from the external world. While dreaming sleep differs from waking consciousness, when levels of relevant brain activity are measured with a metric called the Perturbational Complexity index, dreaming is far closer to waking consciousness than other phases of sleep, or being in a coma, in a vegetative state or under anaesthetic.

Sleep is also essential to the majority of species. Mammal, bird and even fish health deteriorates sharply with a prolonged lack of sleep. All of these animal groups go through alternating patterns of REM (rapid eye movement) and NREM (non-REM) sleep, as do we. There are some significant differences between the nature of sleep in different species, of course. For example, in birds, REM and NREM patterns are much shorter than in mammals. Like conscious breathing whales and dolphins, birds can also send one half of their brains to sleep at a time. However, even allowing for these differences, the scientific consensus is that other animals do dream.

One apparent purpose for dreaming is to rehearse real-life experiences. In a 2010 study, participants were given a set number of hours over three days to practise an arcade skiing game, Alpine Racer II.[45] The on-screen avatar's movements are controlled by the player, who stands on pads that move with a similar motion to real skis, giving a motor-learning dimension to the game. A sleep monitoring programme woke participants up during their early NREM phases, the time they were most likely to recall their dreams. In these NREM phases, unsurprisingly many reported dreaming about skiing when the dream was interrupted. One reported particularly vivid dreams, re-running a tricky corner in the game.

Although we cannot communicate with animals about their private worlds, we might be able to predict the content of their dreams from observations of brain activity, because we know that dreaming brain activity is similar to waking brain activity, for similar actions.

For example, the male zebra finch spends many hours perfecting its song, to maximise its chances of attracting a mate. Two studies were able to match neuronal activity in male zebra finches with similar neuronal patterns as the birds sang during the day.[46] This matching of brain activity in sleeping and waking suggests the animal's experience would be essentially the same. Slight variations in the detected patterns meant the finches might have been experimenting with variations of their song in their dreams.

Similarly, in 2000, another study trained rats to run on tracks and through mazes in order to receive a food reward. The rats had tiny monitors implanted in their brains, which revealed that there were distinct patterns of neural activity at different parts of the course. The researchers then monitored the neural activity in the rats' brains as they slept and found patterns that matched the learning sessions on the track, both during REM and non-REM sleep.[47] This suggests the rats replay and re-experience their daytime activity.

Given such studies, and assuming a continuity between species, it would be odd if other animals' dreams did not serve a similar function to our own in terms of sorting and consolidating memory, facilitating learning, and preparing for future events. Many animal sleep studies are, after all, conducted with the intention that they might tell us something about the role of dreaming and human memory, which assumes a continuity between species separated by millions of years of evolution.

Dreams are of course not definitive proof of subjective experience. We can imagine the songbird has no internal experience of its song, and the rat no internal experience of the track. It is *possible* that there is nothing *it is like* to be a running rat or a singing zebra finch, either in dreaming or waking. Perhaps all we are observing is a repeated process, like a computer that periodically de-fragments its hard drive, and goes over memories while it reorganises the data, while lacking any first-person experience.

Yet it seems a profoundly odd idea that across species, evolution has led to similar mechanisms in brain activity, similar patterns of sleep and waking across species, a similar physiological need for sleep and a night-time replaying of daytime brain activity, yet after half a billion years of vertebrate evolution, we are the only species with any first-person experience, in dreams or during waking hours. If we accept that other animals do dream, they are also likely having some sort of first-person experience: cats dreaming of catching mice; dogs dreaming of rolling in stinking puddles without being sent for a bath afterwards; bears dreaming of finding the perfect stash of honey. First-person experience is what dreams are made of in humans, and as other animals dream, at least some of the time they too are likely experiencing a singular point of view.

Those who strictly follow Morgan's parsimony, and are sceptical of other animals having meaningful first-person experience, tend to argue that no one really believes animals are black boxes of perception, like Descartes' biological machines. They just don't believe there is enough clear evidence to conclude other animals are having a conscious experience.

However, it would be wrong to insist that first-person experience must remain a human-only attribute, when the evidence shows the brain activity of other species in waking and dreaming follows similar patterns to our own.

The difference—and the reason the playing field is so heavily tilted in our favour—is that we are able to validate internal experiences between ourselves through a common language. Behaviourism was historically hostile to trusting first-person accounts of anything, and it has been said that a *true* behaviourist would deny that even people are capable of first-person experience! Having moved past behaviourism, we need to make an allowance for other species' lack of language, and be willing to make some predictions about their lived experience.

The brainstem theory of consciousness

One neuroscientist who thinks lived experience is key to answering the question of consciousness is Mark Solms.

When Solms was an undergraduate in South Africa in the early 1980s he asked his lecturers about consciousness, about where the subjective "I" would be found in the brain. Not only did neuroscience not have an answer at the time, the young Solms was cautioned that such questions could even be damaging to his career.

However, greater willingness to address this issue came in the early 1990s, and the consensus for a location for consciousness settled on the cortex, in particular the frontal cortex.

The frontal cortex is the region that carries out many of our "higher functions", such as language, reasoning, planning and social awareness. Given that, at the time, the consensus was that only human beings were truly conscious, it seemed obvious that the place all the sensory input came together to create conscious experience would be the known location of our species' most impressive cognitive abilities.

In his 2021 book *The Hidden Spring*, Solms presents evidence in favour of it being not the cortex, but rather the *brainstem* that brings our subjective experience together. Take a moment to consider how radical an idea this is. The cortex is, by mass, about 40% of the wrinkly walnut-shaped organ which fills our skulls.[48] Because this outer layer of the brain includes the functions associated with human intelligence, it was considered the most likely home for consciousness. Yet Solms is one of a growing number of scientists who believe consciousness resides in the brainstem. The brainstem hangs off the lower part of the brain, and was long assumed to be little more than an interface to the body's nervous system.

Solms addresses the metaphysical issues for neuroscience directly, devoting a whole chapter to Chalmers' hard problem. He comes down squarely on the side of subjective experience, namely Chalmers' *blueness of the blue*, or Nagel's *what it is like to be*, as the thing we should identify as consciousness (rather than cognitive abilities). Solms has a simple point about subjective experience being the key attribute of consciousness, in that subjective feelings cannot be experienced unconsciously. It is not possible to experience colour subjectively, or have any sense of self while under anaesthetic or in a coma.

Further, being awake is not necessarily the same as being conscious of everything around us. Consciousness is highly selective. We routinely perform learned tasks like driving, walking, opening doors, with minimal conscious attention. Our subjective experience of the world is made up of those things that need to occupy our attention at that moment. As an example, put yourself in the steel-toe capped boots of a construction worker on site, installing a heavy joist while there are workmates down below whose safety demands your attention.

Once the joist is in place, you might then realise that during the installation you had hit the edge of your thumb with a hammer, but ignored the growing pain until the danger to the people below had passed, because the brain prioritises what it will allow to fill your conscious attention.

If consciousness is defined by the ability to have a first-person experience, rather than the ability to perform tasks (in other words, the opposite of one of Descartes' unfeeling and unknowing biological machines) then experience generated by the brainstem would be a radical departure from the inherited idea that consciousness depends on the most complex structure in the primate brain, the cortex. In evolutionary terms the brainstem is about as primitive a part of the brain as it gets, as it is found in all vertebrates. This potentially locates the dawn of animal consciousness with the appearance of the first vertebrates over half a billion years ago, rather than the evolutionary blink-of-an-eye since humans started making tools and building civilisations.

Solms' case for locating consciousness in the brainstem is more than deductive, however. It is backed up by clinical evidence. There are many individuals with damaged or missing areas of their cortex, who, although they have impairment to specific cognitive functions, are still partly or fully conscious. Whereas damage to an area as small as two cubic millimetres in the brainstem appears to be enough to switch off consciousness completely.

Most striking perhaps are Solms' own case studies. One of his middle-aged patients lost the majority of his frontal cortex as a teenager. Fortunately however, his Broca's brain area (a region crucial for the production of language) was left intact, so he could answer questions, reflect on his own experience, and even crack a joke when given a test requiring some imagination. Patient W, as he is known, is able to provide a rare subjective account of first-person experience in someone who is missing a very large part of his frontal cortex. There is no more reason to doubt Solms' patient W when he claims to have a sense of self than any person with a complete and healthy cortex who tells us the same. Taking Solms' patient at his word, the frontal cortex does not appear to be essential to his first-person experience, or a sense of self.

The insular cortex in the middle of the brain was another candidate for being the source of consciousness. But once again, Solms is able to relate testimony that contradicts this. A man known as patient B from Italy lacks insular cortex because it was destroyed by a virus. But in interviews, Patient B makes as convincing a case for being self-aware as you or I would. In essence, whatever role these regions play, such accounts suggest consciousness is not wholly reliant on them.

Accounts of children with hydraencephaly are even more striking. This rare condition sadly means a short life expectancy. During pregnancy the cortex fails to develop, leaving only a brainstem, and the skull instead fills with cerebrospinal fluid. However, children with hydraencephaly are in many ways like other children with a severe disability. They sleep at night and wake up in the morning. Many have vision, hearing and tactile senses, and their parents report clear likes and dislikes for toys and cartoons. In *The Hidden Spring*, Solms describes five hydraencephalic children and their families taking a trip to Disneyland, which was clearly loved by all. The children were delighted to meet Mickey Mouse, had favourite rides, got overly tired and cried sometimes. They essentially behaved as many other young children would at Disneyland.

For those who believe consciousness is the product of a complex brain structure, the uncomfortable yet inevitable question would then be, are these childrens' behaviours only simple reflex reactions? Can there really be an individual experiencing the world through sight, hearing and touch if there is no cortex? Unlike Patient W, because the children are unable to speak, they cannot describe their inner worlds to anyone, so again, a degree of supposition is required.

Solms has little doubt these children have first-person experience and should be considered conscious. This is not just a matter of opinion, as there is an independent measure for the level of consciousness. For clinical and ethical reasons, patients in comas are rated on the Glasgow coma scale, which rates eye movement, verbal and motor skills to give a maximum of 15 points, a top score being fully conscious. On this objective measure, hydraencephalic children would score near the top, which is far higher than individuals with a complete cortex and brainstem who are in a vegetative state or induced coma.

Here Solms warns of the ethical danger of being so wedded to the cortex as the location of consciousness that someone responding as if they were conscious, as seen with hydraencephaly, is considered insufficient evidence that they actually *are* conscious. He writes:

> Let me therefore put this point forcefully: if we are to accept that someone who seems to be conscious actually isn't, we should require an extremely convincing argument. Merely raising philosophical doubts isn't enough. We need very good grounds to think that the two sorts of consciousness [wakefulness and phenomenal consciousness] have come apart in such people, as they seemingly never do in us.[49]

Given the cause of hydraencephaly, it's reasonable to ask whether someone with this condition has perhaps retained a small amount of cortex. Perhaps some cortex became part of the brainstem early on in pregnancy, but is no longer detectable?

This question was addressed experimentally by the neuroscientist Jaak Panksepp, who surgically removed the entire cortex of a group of rats. He found that even without any cortex the rats still continued to act like rats. They swam, played, ate and were capable of being both territorial, and the females maternal. When Panksepp asked his students which of two juvenile males they believed had a complete cortex, most chose the rat *without* cortex, because he appeared to be the more active of the two. The rat *without* cortex was judged to be the more rat-like!

For now, set aside the moral implications of Solms' brainstem hypothesis, and leave out the very real question of rights, both human and animal that arise from it. Take this information about brainstem consciousness only for what it implies about our concept of a conscious being. If people can be deemed aware and conscious without a cortex (as it appears they should be) and a rat can continue as a functioning social being without any cortex at all, then there is a strong possibility all other vertebrates have some first-person experience, because the size and complexity of their cortex is less an indicator of whether they are conscious than the fact that they have a brainstem. It follows, there likely is a *what it is like* to be a bat and a *what it is like* to be a grizzly bear. There could well be a *what it is like* to be a fruit fly.

Bats and bears are clearly distressed by pain, they dream, and from one neuroscientist's point of view, there is reason to believe someone is "at home", because they possess this basic part of the brain's hardware. The reluctance to allow other species to have subjective experience is mostly due to the lack of a common language, language that would give us a description of their inner worlds.[50]

The brainstem hypothesis means we should attribute consciousness to the vast majority of animals. The brainstem hypothesis, however, would not recognise trees as conscious, because they lack a brainstem. If the minimum requirement for consciousness is having a brainstem, consciousness probably emerged some half a billion years ago with the first vertebrates. Solms is not alone in putting the dawn of consciousness so far back. Evolutionary biologist Eva Jablonka, and neurobiologist Simona Ginsburg argue that Unlimited Associative Learning, or UAL, should be taken as the marker of consciousness.

There isn't space to describe UAL properly here, but it is worth noting that Jablonka and Ginsburg drew their approach to the evolution of consciousness from principles established in evolutionary theory.

The key to lifeforms is that not only can they replicate, they can also give rise to unlimited other forms of life, which are limited only by the niches they evolve to occupy. In a similar way, mental capacities of species are unlimited, in that the most minimally conscious form of life can, through evolution, give rise to more complicated consciousness. For this reason, Ginsburg and Jablonka put the dawn of consciousness at the Cambrian explosion, 540 million years ago, when life developed into the major groups (animals, plants, fungi, bacteria etc.) that we know today. This is when brainstems first made an appearance.

In fact, even if we decide that consciousness depends on the pre-frontal cortex, some animal studies suggest it could have started some 300 million years ago with the ancestors of corvids (the crow family). It is a mark of how the consensus has changed in this respect that the abstract of a 2020 University of Tübingen study, which showed similarities in prefrontal cortex structure and activity between crows and primates, was able to state their study "demonstrated for the first time that corvid songbirds possess subjective experience".[51]

They are asserting that crows should be considered *conscious*, which perhaps 10, certainly 20 years earlier, would have been considered a controversial statement. Of course, if there is an element of mind in all things, that would require a different take on any potential dawn of consciousness. Whether it emerged over 500 million years ago in early vertebrates—as UAL and the brainstem hypothesis suggest— or at some point since our ancestors started walking on two legs around 4 million years ago, the metaphysical issue of how life might spontaneously have become aware enough to be capable of having a first-person experience remains.

Whether we decide it began in the human cortex relatively recently, or in the brainstem half a billion years ago, there is still this question of *emergence*. Here again I raise the ability of the no-brain (and no-brainstem) slime mould to navigate mazes, map out transport networks and pick a healthy diet, in questioning what externally verifiable differences there are between those beings deemed conscious, and those deemed to have the lesser status of being simply aware of their circumstances.

Comparing cognition

Having considered experience, let's return to the other facet of consciousness, cognition, and the long-held assumption that true consciousness depends on a sufficiently complex brain structure.

In humans, an individual's level of intelligence or skills does not determine the rights they are afforded. More intelligence or a larger number of competences has advantages for the individual but does not lead to more rights or privileges in law—the dangerous notion that societies should be built in this way led humanity down the dark roads of eugenics and forced sterilisation in the early 20th century. As with computers, we can certainly conceive of intelligent behaviour with limited or no understanding, of there being competences without comprehension. Extra competences do not help us rank consciousness for members of our own species, so should they between the species?

For now, let us entertain the idea that true consciousness is not only the first-person experience in the brainstem, it must also be combined with certain competences.

Neuroscience tells us that much of what we sense is processed subconsciously. To allow us to function, the brain does a considerable amount of work that we are never aware of. Some "higher order" theories of consciousness argue that our perceptions only become truly conscious when a more sophisticated process, such as language or reflection is involved.[52] In this way, these extra competences transform sensing into a first-person experience, by adding the element of reason, or reflection.

However, when competences are taken away from humans, their rights and status as conscious beings usually does not change. Take planning. Damage to the frontal lobes can cause a loss of executive function, including the ability to plan. We can, of course, find evidence of planning in other species. A chimp named Santino in a zoo in Sweden often built a stockpile of stones, that he would throw at visitors. This suggests both a theory of mind (knowing that his keepers would take the stones away if they were not hidden) but also the capacity to plan by collecting them together in a location for future use, ready for when someone passes by who he particularly dislikes. If we say consciousness depends on having a certain level of executive function, then evidence of planning in a chimp should elevate its consciousness.

Returning to Solms' Patient W, his having no frontal lobes is a severe impairment that profoundly affects his executive function. He was living without the part of the brain that, according to consensus, should be responsible for true consciousness. But when interviewed it became apparent he was still able to reflect on his own existence and had a clear sense of himself. So was Patient W mistaken? Does the loss of

171

certain competences make someone a philosophical zombie, able to spout words without true conscious experience? This seems very unlikely.

Adult traits such as self-recognition and language take several years to develop, so it might be argued that true consciousness is not present at birth, and is in fact only attained after years of learning and development. If other animals need sophisticated cognition to be truly conscious, we might wonder, using the same criteria, would a human baby meet the threshold for consciousness?

What *is it like to be* a baby? We were all there once, going through the all-consuming processes of learning to become a person, but none of us can remember those experiences. One reason most people have few, if any, memories from before kindergarten is that the young brain is forming neural pathways that give us general rules for living. This generalisation allows us to perform many tasks, because unconscious processing is more responsive, and requires less energy than conscious processing.

As every parent knows, childproof catches are a kitchen essential, as toddlers love opening cupboard doors. Yet I doubt any of us can remember when we first figured out how to open a door. The young brain is unlikely to be storing each instance of door opening into a memory for future reference. Instead, the brain gave the task as much attention as it needed, until it was able to transform the practice of reaching out for a handle and pulling a door open into a general rule. This generalisation of actions may be the reason why a newborn's sleep is made up of 50% REM sleep, compared to an adult's 20-25%. A baby's brain likely needs more of that type of sleep to determine what from the day's activities can be turned into generalised rules, that can be applied in future to different contexts.

Returning to scepticism about animal experience, if animals have a similar level of consciousness to an infant, perhaps what happens in the animal mind is more like the experience of a human baby? Maybe their mental worlds are a kind of black box of sensations and responses that are never properly ordered, because their brains never acquire key skills, such as language, that would allow for the highest level of abstraction, awareness of self and the organisation of one's own thoughts?

Again, the behavioural evidence does not suggest that the baby and non-human consciousness are especially analogous. One reason human babies are born with such undeveloped brains is our bipedal physiology. There simply isn't space for a larger newborn, with a brain ready for infant life from day one, to pass through the birth canal. By contrast, many herd animals have only minutes to get to their feet after they are born. Almost immediately they must see, hear, smell, run, vocalise, feed from their mothers, and keep up with the herd as best they can.

Take one of the smartest herd animals, dolphins.[53] Within minutes of emerging from the womb, a dolphin calf must swim, be aware of its marine environment, and recognise its mother's signature whistle, or be at great risk from predators. As previously mentioned, dolphins are without a doubt conscious in one vital respect from birth—they are conscious breathers. Every breath a dolphin takes throughout its lifetime is a conscious act, and any dolphin that loses consciousness for long enough will drown. For this reason, the newborn dolphin is pushed to the surface by its mother immediately after birth, so it can take its first breath.

In a 2018 study, two young dolphins, one seven months, the other sixteen months old, seemed to recognise themselves in a mirror.[54] By contrast, a human baby takes at least a couple of years to reach this point. Unlike the young dolphin which can move more or less like an adult from birth, human babies only start moving themselves around by crawling from around six months old.

Whether or not one accepts the mirror test as evidence of self-awareness, it is clear from the newborn dolphin's behaviour that whatever level of consciousness an adult dolphin has, the newborn dolphin is far closer to its eventual adult state than the human baby is to adult consciousness. This makes the notion that other animals live in a mess of sensory inputs, responding with little or no comprehension like a human baby, very unlikely. If we predict inner experience on the basis of neurology and behaviour, because there is evidence for some abstraction (an ability to weigh up options, plan ahead, have a theory of mind etc.) in non-human species, this puts their consciousness closer to adult humans than the infant human. Again, one could argue that in the case of animals these may be competences without comprehension.

But if we treat increasing competences as milestones on the road to the baby becoming a fully conscious adult, the existence of competences in young and old animals should also be counted as indicators of their level of consciousness. Thanks to our unique evolutionary position, our species is afforded the time to develop into adult consciousness. But it does not follow that other species are forever stuck in a fuzzy baby-like mental space.

Of course, no thinkers are actively promoting the idea that animal experience is the *same* as the human baby's experience. Given that, you may wonder why I have spent a couple of pages on the subject. The reason is, as I said earlier, if we only look at behavioural evidence without then asserting some idea of what it might be like to be a grizzly bear or a dolphin, if we treat other animals' experience as a wholly inaccessible black box, we are left with looking to something that does not really fit the bill, such as *what it is like to be a baby?* as a rough analogy for animal experience, solely because it is a lower form of awareness we can relate to, and imagine. If animal behaviour is not down to sub-conscious generalised processes that require no thought, one important piece of evidence in non-humans would be experimental evidence for distinct patterns of conscious and unconscious processing in an animal brain.

To test conscious and unconscious vision, macaque monkeys and humans were set essentially the same test. Participants had to predict on which side of a screen an image would appear by pressing a left or right button. When repeatedly given a brief, but consciously visible cue on the opposite side to where the image was about to appear, both humans and macaques soon learned that there was a pattern, and chose the opposite side.[55] When the visual cue was flashed up too quickly to be registered by the conscious mind, both humans and monkeys tended to pick the side of the misleading cue, meaning they registered the cue subconsciously, but were unable to take enough of a step-back to choose the opposite button.

The fact that there is a difference in reaction to the short and long cues, suggests that the monkeys, like us, have a conscious visual field and an unconscious one. We know from behavioural evidence that other primates have moral behaviour, tool use, theory of mind, and gesture language. The ability to overrule an impulse for a more strategic purpose, is considered evidence of consciousness. As humans and monkeys perform the task in the same way, the existence of a conscious visual field indicates that a higher-order process exists in monkeys. Given enough time, the monkey brain, like the human brain, seems to have the capability to step-back from the visual information entering its brain, and change its behaviour if required. This means the monkey is not just experiencing, it is pausing to consider and evaluate, to *think* about that experience in some way.

A comment on free-will

A shift away from a consensus that the cortex is the home of consciousness has significance for another long-standing question around human existence, the question of whether we have free-will. Although I am not looking to make a pronouncement on one of the oldest philosophical debates, it is worth pointing out neuroscientific arguments against free-will are based on what is already an out-of-date hierarchical concept of human consciousness.

Any number of prominent scientists will state that a belief in free-will is incorrect, because science has proven every action we take is pre-determined, meaning consciousness can only be a witness to our actions, and any sense of control we may have over those actions is therefore an illusion.

Broadly speaking, there are two parts to this claim. From a cosmic standpoint, science says everything that happens right now is the end product of a chain of cause-and-effect events, stretching right back to the big bang. Further, the composition of a brain, which generates our consciousness and appears to make decisions, is also the product of these causes and effects. Therefore, the scientists argue that a belief in free-will would require some non-physical self—a *real me*—to sit inside our heads and pilot the ship, ready to influence the chemistry and electrical signals in our brains to produce different behaviours. In this sense, a lack of free-will is consistent with materialism and physicalism.

Ironically though, scientists who take this determinist view are rather undermining the basis of the entire Western scientific method. After all, experiments are conducted under different conditions to test the outcomes produced with known variables. We conduct experiments where we choose different variables, to test different outcomes—all of which is rather meaningless if everything in the universe is wholly determined, because that would mean experimental outcomes themselves are all pre-destined.

But I am not looking to debate whether cause and effect in physics means determinism is correct. What is more interesting is how neuroscientific experiments into free-will, of the type begun by Benjamin Libet in the early 1980s at the

University of California, are often cited as biological evidence in support of this determinism.

Libet set volunteers tasks with a freely chosen moment of action, as an EEG machine monitored their brain activity. Volunteers were instructed either to press a button or flex their wrist, an action to be performed whenever the subject wished. They would perform around 40 repetitions of this freely chosen movement, while also logging the moment they decided to act by noting the position of a fast moving spot on a clock-face. The dot made one 360 degree revolution every three seconds, so by noting the position of the dot, it would record the moment of decision to within fractions of a second (roughly 50 milliseconds per mark). The EEG meanwhile was monitoring what Libet called the "readiness potential", namely the electrical activity in the motor cortex that precedes a voluntary muscle movement like flexing a wrist or pressing a button.

What appeared groundbreaking was that Libet's experiment found this readiness potential occurred far earlier than the subject's apparent decision to move. Counting back from the actual moment the wrist moved or button was pressed, the average readiness potential was approximately 550 milliseconds earlier than pressing the button or flexing the wrist.

But the position on the dial observed by the subjects seemed to indicate that the conscious part of their decision came much later, only around 200 milliseconds before the actual movement of the wrist or finger. This suggested that something in the subconscious, down in the motor cortex, could be choosing the moment to flex the wrist or press the button, not the conscious mind.

From this, many have concluded that, what for the subject felt like a moment of conscious decision making, was really the delayed report of an action led by the motor cortex. As movement starts before people are strictly speaking, *conscious* of the action, many believe it proves there is no such thing as free-will. Later experiments have also shown that in situations where we have less control, our confidence in our degree of agency goes up, not down.

One experiment at University College London has similarities to the one with macaque monkeys described in the previous section. Participants were asked to watch a screen and press one of two buttons in response to a left or right pointing arrow: the left button for a left arrow, the right button for a right.

The element of choice was brought into the experiment by sometimes presenting a double-headed arrow, for which participants chose either the left or right button.

What they had not been told, however, was that sometimes a small arrow pointing either left or right would be flashed up on the screen, for a fraction of a second before the double-headed arrow appeared. This arrow appeared so fleetingly it could not be registered consciously. As had been found with macaques performing a similar task, when the subliminal suggestion was present, people were more likely to press the button that matched the subconscious cue.

When asked to rate how strongly they felt they had been in control of their decision making, there was a small but statistically significant increase in their sense of control—a roughly 5% higher level of confidence—when they had in fact been unknowingly prompted. In this case it appears that the sense of being in control of our actions increased slightly when the subconscious had been primed to take one option over another.

What is supposedly shocking about these free-will experiments is that the more "primitive" part of the brain, the bit closer to our motor regions of the brain, seems to be leading decision making in certain circumstances. Does this mean free-will does not exist? Or that free-will is not what we thought it was?

Experiments like Libet's still hold considerable sway. But critics have raised several issues with the original experiments. First, the timings I gave earlier—readiness potential at 550 milliseconds before movement, followed by conscious awareness at 200 milliseconds before action—are averages from many runs of the experiment.

Among the volunteers there was a range of intervals between the readiness potential appearing, the subject being aware of their intention, and the actual physical movement. Later experiments that used auditory cues have shown a considerably shorter interval between the readiness potential, and the volunteer's sense of having made a decision.

Second, measuring the moment of decision by observing a dot on a dial, requires the volunteer to report on the moment of their own volition. So should we subtract 70 milliseconds or so from those timings, due to the processing delay of visual information as a result of chronostasis, described earlier in this chapter?

Third, a more esoteric problem with this method is, were the volunteers reporting the actual moment they made their decision? After all, their attention is divided between performing the task and observing their own mental state. Could the long delay be partly explained by their having to report the moment they became *aware of becoming aware* of their own will?

Fourth, the readiness potential itself is not universally accepted as evidence of an action being performed by the subconscious. It may only be showing that the nervous system is preparing for movement, a sort of primer, which gets the muscles ready for the conscious decision which actually turns that decision into movement.

Lastly, there is a question about what these experiments are actually testing. They often involve basic physical movements, flexing wrists or pressing buttons. To maximise our chances of survival, as hunter gatherers we are generally hard-wired to react first, reflect and analyse later. But those taking part in Libet's original experiment were not reacting to a stimulus. Instead they were given an open-ended period of time to choose the moment of action, while using a part of the nervous system that has arguably evolved for *re-action*, rather than a freely chosen moment of action.

It is worth noting that despite his ground-breaking experiments, Libet himself did not believe in a fully determined universe.

His later experiments studied the power of veto. Participants were again instructed to intend to move their wrists or press the button, but then to immediately veto this action. This experiment, in particular, led Libet to believe that free-will is primarily the power to veto an action that starts with the subconscious, a view that is often characterised as *free-won't* rather than *free-will*.

How does this question of free-will relate to changing views of the brain? For much of the 20th century there was an assumption that human conscious awareness must be located in the cortex or the frontal lobes. The frontal lobes in particular are often called the CEO of the brain as they are said to perform the executive functions that involve rational decision making, when different sources of information are brought together.

Executive functions give us the ability to plan and to consider the consequences of our actions, rather than simply following our immediate impulses. For example, no matter how strong an attraction we might feel, we do not grab strangers in a lift and kiss them, because we have enough of a theory of mind to know the object of attraction could be distressed by this, and that there would be social and legal consequences from such an action.[56]

Because this CEO was supposedly the ultimate authority over all our important decisions, the same brain region was assumed to be the home of consciousness. However, while the idea that consciousness may in fact reside in the brainstem, not the cortex, argued for by Mark Solms and others, is comparatively new, if correct it greatly undermines that assumption. It undermines the conclusion, from the button-pressing type of experiments at least, that free-will is an illusion, because the frontal cortex is likely not the seat of consciousness.

Awareness, self-awareness or sentience?

The current scientific consensus seems to be that consciousness is not determined by one key trait. It is not just language, or planning, or problem solving, or anything else that unlocks the door to a special mental space, labelled "true consciousness". The idea of a single thing making humans unique is, to me, more a religious than a scientific impulse (which I will look at in the next chapter). Science now accepts that other species are experiencing the world through more than a simple behaviourist stimulus/response model. Traits which may suggest consciousness include: raised heart rate under stress (indicating a self, because the self is currently under threat); flexibility in decision making; some theory of mind; a requirement for sleep; having a greatly altered mental state under anaesthetics.[57]

I won't try to assess all these traits here. But it is worth briefly considering research on one—the extent of focused attention in other species. What ability do they have to focus on one thing from a range of competing stimuli?

In the immediate moment, we receive an enormous amount of information from our senses as they compete for our attention. In order to avoid being a piece of flotsam on a sea of stimulation, the mind must be capable of focusing on the most important information, and push other information into the background. For example, the construction worker focusing on safely installing the beam while temporarily ignoring the hit thumb. It is apparently common sense that directing attention to something specific would be essential for most animals' survival.

But focused attention also requires a choice to be made between competing stimuli, which implies someone is "at home" to *make* that choice, and focussed attention is something that even the humble fruit fly appears to possess.

Researchers have implanted probes in the centre of fly brains to listen in on neural activity. When the fly heads in the direction of an object of interest, the brain waves make a distinct pattern that has similarities to the way another animal brain behaves when directing its own conscious attention—the human brain. They also found flies lose interest in a new shape displayed on a screen after a time, which suggests that even a fruit fly becomes familiar with new objects, and perhaps can even become bored.[58]

Neuroscience is showing that even the simplest animals may be having some sort of first-person experience. In favour of this conclusion is the fact that they have brainstems, they pass in and out of dreams just as humans do, and they selectively adopt a point of view for their own survival. So here again, we come up against a philosophical split. On the one side are those who believe that consciousness is defined primarily by the ability to have an experience. To this, a brain may add cognitive skills. On this side of the debate, the word *conscious* could apply to a very large number of species.

On the other side are those who believe that consciousness requires a brain equipped with higher-order abilities, which transform the flow of sensory inputs and outputs into true consciousness.

The scientific consensus in the last decade seems to have been moving towards the first. For example, Anil Seth said this in an NPR interview in 2018.[59]

Consciousness is any kind of experience at all, whether it's a visual experience of the world around us, whether it's an emotional experience of feeling sad or jealousy or happy or excited. From experiences of intending to do something or being the cause of something that happens, consciousness is the word that we use to circumscribe all the different kinds of experiences that we can have.

Because human society requires us to spend a great deal of time being self-conscious, aware of our personal space, knowing when or what is an appropriate thing to say in a specific situation etc, *true* consciousness is often said to involve a high degree of self-awareness.

As mentioned in Chapter 2, in 1970 Gordon Gallup devised the mirror test for chimpanzees to determine whether they are self-aware. While only our closest primate relatives passed the mirror test, it was generally accepted as an indicator of a self-aware consciousness.

The value of this test is currently being re-assessed, however, because one study found that a small fish, the bluestreak cleaner wrasse, could qualify for a pass.[60] This fish, which makes its living by offering cleaning services to large predators, routinely swims into sharks' mouths. It would seem to have more reason than other fish species to anticipate the behaviour of any larger animals it comes across, and it would seem beneficial for it to have an awareness of how it might appear to other animals. Even so, many do not accept that the cleaner wrasse has passed the mirror test.

Without going into the arguments for and against the bluestreak cleaner wrasse being considered self-aware, the findings did not fit with the dominant assumption that self-awareness requires a large, highly complex brain. But if we accept that the cleaner wrasse has some self-awareness, then there is no stepping-up of brain size and structure that results in a clear *switching on* of self-awareness at some point. If the mirror test is evidence of self-awareness, as neural complexity increases from the brain of a cleaner wrasse to that of a chimpanzee, there are still few animals, fish, birds or mammals that pass the test.

Self-awareness is vital to our survival. We also find behaviours that indicate awareness throughout the plant and animal kingdom. Yet as we work down the tree of life, it becomes increasingly difficult to determine whether that awareness might include a degree of self-awareness. I have no idea if slime mould is aware of its own oozy yellowness, or experiences the *oatiness of the oatflake*—in such simple forms of life comparisons with larger animal biology will never be possible. However, as outlined above, there are reasons to believe fruit flies, dolphins, bears and most other species, could also be living through their own private cinemas of experience, to some extent.

Self-awareness exists on a continuum. Even in humans, self-awareness is not considered to be a binary process. When a two year old first recognises itself in the mirror, it has not made a chrysalis-to-butterfly type transition from a world of simple present-moment awareness to wondrous self-awareness. The self-aware social being it becomes is the result of many more years of learning and development. Is a bacterium self-aware when it moves away from a toxin or toward some food? It is acting in its own self interest. Perhaps it does have some awareness of itself as an entity, even if that is only one thousandth of the awareness that helps it to arrive at a course of action?

We may even want to revisit the terminology, and say the brainstem hypothesis indicates not so much that the majority of animals are *conscious*, but are likely having an experience and so are *sentient*.

We might then reserve the word *conscious* for those animals we have reason to believe have both a conscious and subconscious stream of processing. This would not necessarily downgrade other animals' mental worlds. Rather, I would suggest we downgrade the term "conscious" from its current lofty position, because its cumbersome and mediated way of interacting with our environment requires a subconscious stream to *support* the conscious stream. Other lifeforms may be managing perfectly well without this dual stream, and experiencing life while being highly aware of what is going on around them, and their own possible responses.

Arguing that there is no clear line to be drawn between the conscious, the sentient and the aware, might be seen as sitting on the fence. This is not due to a lack of willingness to engage with the evidence. It is because as the neurological evidence grows, the picture of neural sophistication that emerges is unlikely to have many sharp edges. This is not necessarily a problem, as the boundaries that differentiate one thing from another can be fuzzy, even in science. For example, in evolution there are not always clear lines to be drawn between species as a new species evolves. Instead there are *transition markers*, signifying the beginning of a change that in retrospect turns out to be significant.

We should see consciousness, awareness and sentience working in the same way. Science should be doing the work required to have more granular definitions about all animal cognition and experience. But we should also realise that any dividing line will be less pronounced when we recognise that the fully-developed adult human consciousness, with the ability to examine its own thoughts, is simply one subset of animal awareness. This animal awareness could exist because there is an element of mind in all things, rather than being the end product of a specific form of neural scaffolding.

7

The myth of separateness and superiority

The opening sequence of Stanley Kubrik's film *2001 A Space Odyssey* is titled "The dawn of man". As the Sun rises over the African plains a group of early primates wake to find an enigmatic black obelisk in their territory. At first they try to scare it by shrieking, beating their chests and waving their arms. But curiosity takes over and they move closer until one bold individual places a hand on its smooth, dark, other-worldly perfection. Soon the others follow suit. Later that day, their leader picks up a large animal bone and delights in smashing a skeleton to bits. His troupe starts using bones for hunting, so that instead of scavenging for a living they are now effective predators. Armed with these weapons, they attack a rival group of apes, beating one to death and gaining control over a water hole.

After their victory, one ape hurls his weapon up into the sky, and as the bone drops in slow motion, Kubrick delivers one of the most iconic jump-cuts in cinema history—the falling leg bone becomes a space ship falling through space, far above Earth. The conceptual leap is clear from this sudden shift of time and location. The obelisk has sparked off something in the minds of our ancestors that leads to tool use and abstract thought. This makes primate technology possible for the first time, and puts the human race on a singular path, ending in the modern human story and space travel, six million years later.

Whether or not Arthur C. Clarke believed an alien visit actually have jump-started human development is unimportant. The obelisk is largely a narrative device. The reason this 14-minute sequence with no dialogue has power for the audience, is because it taps into a common belief that we are the only species to have made the leap to an elevated mental state. In Western culture, the leap may be attributed to extra-terrestrials, or through divine intervention represented by the finger tip of God in Michelangelo's *The Creation of Adam*, or more gradually after evolution created some kind of consciousness switch that flipped itself to the *on* position in our brains alone. Whatever the apparent cause, we understand the opening of *2001* because there is a pre-existing narrative that humans belong in a fundamentally different category from all the other species.

When human consciousness is discussed in popular science books and videos, some well-worn phrases often appear. Human experience is often described as *rich, lucid, wondrous, enigmatic,* or *vivid.* On social media, memes tell us the most complex object in the known universe is the organ that sits on top of our shoulders, a message inscribed over an image of the cosmos.[61] Equally, the phrase "the mystery of consciousness…" is rather overused, and usually delivered with a sense of awe and reverence.

In one sense, mystery is a valid term for consciousness. A mystery can simply refer to an unsolved problem. But the word has other connotations and is frequently invoked to signal something as magical and beyond the rational mind, such as the *mystery of the resurrection of Christ.* Similarly, our secular culture sometimes treats human consciousness with a near-religious awe, that almost places it beyond mundane natural forces. How our culture creates and reinforces this assumption, this myth, needs to be addressed, if an element of mind in all things is to become a less obscure philosophical idea.

Culture is no longer considered unique to humans. Meerkats can be said to have culture, as one group may get an early night and rise early, while another group stays up late and rises late, even though the environmental pressures on the two groups are essentially the same.

There are killer whale groups with overlapping territories that appear to have distinctly different dialects. They also have different feeding habits—one group will only feed on fish, another group will hunt seals and other marine mammals.

Culture deserves a chapter of its own because, while there are instances of culture in other species, the *extent* of human culture is undoubtedly unique to our species. Our lives are constantly filtered through culture, and more of our attention can be devoted to ideas and concepts than objects that exist in physical reality. The extent of human culture means our worst traits also persist through the generations—our prejudice, hatreds, ignorance and division.

As language is the primary means of developing and spreading culture, I want to first look at some cultural theory and linguistics, in order to critique some of the stories, the *myths*, that our culture recycles. Our understanding of the relationship between humans and the rest of nature is built around these myths.

It has been said that linguistics professor Ferdinand de Saussure was as influential as Freud and Marx in the development of early 20th century thought,[62] although Saussure himself died before he could enjoy any such plaudits. He gave a series of lectures at the University of Geneva from 1907 to 1911 that his students posthumously compiled into the book, *A Course in General Linguistics* in 1916.

In a way, this book did for culture what quantum mechanics was doing for physics, as his theories undermined the notion that anything written or said ultimately refers to a fixed, objective reality. His work effectively founded semiotics, the study of signs, and divided these signs, (words and their meanings) into signifiers and signifieds. The signifier is the word itself, for example, sheep. The signified is the concept of a woolly, four-legged creature in our minds.

While that might seem obvious, it is important to understand the signified is the concept of sheep in our minds, not an actual four-legged woolly mammal that goes "baa!". As Saussure says (the italics are in the original text):

> Linguistics then works in the borderlands where the elements of sound and thought combine; *their combination produces a form not a substance.*

Saussure emphasised this distinction because he wanted to make clear that it is the idea of an object, rather than an external object itself, that is being indicated by our words. As evidence for this, he observed that translation between different languages requires an understanding of the intended meaning. He reasoned, if language was simply the business of attaching a label to existing objects, it should be possible to translate freely from a description of the physical world that exists in one language to another. Saussure gives the example of the word *mouton*, which in French applies both to sheep as a living animal and as meat. In English there are two different signifiers: *sheep* for the animal and *mutton* for a specific type of meat that comes from sheep.

Translation is a skill that requires far more than a good dictionary. It requires an understanding of context and the culture that produced the language. A more stark example of this is the way different languages divide up the colour spectrum. The Welsh word *glas* covers a range of colours from blue through green to grey, and has no direct equivalent in English. It might seem that the names of colours in particular *should* describe something entirely natural, something that should function equally in all languages, having an existence independent of any concepts. But colours can be culturally specific. For example, the Ancient Greeks did not think of the sky as blue. It has also been suggested that English did not develop an identity for blue, until woad production in the 13th century made it an affordable colour for garments. It was only then that a specific word was needed to distinguish blue from green, grey or purple.

Saussure also noted that languages vary their pronunciation, and words evolve over time. Again, if language is only the labelling of objective reality, why would this happen if the objects being labelled, particularly those in nature, are not themselves changing?

The lack of direct equivalence also applies to grammar, which varies greatly across languages, and also evolves over time. Perhaps more significantly, different languages rarely have the same number of tenses, meaning that not only do different cultures place different boundaries when describing physical objects, there are also important differences in how they divide up more abstract concepts, such as time.

Saussure argued that this means language should be seen as a system of *difference* not *reference*. So for example, the reason a sheep needs a different word from a goat or a cow, is that despite any similarities these animals are actually being defined by their differences. A sheep is defined partly by its *not being* a goat, or a cow.

Saussure said that language is in many respects a process of exchange, like a currency. Like currency, words can be exchanged for similar things—two 5-franc pieces can be exchanged for a 10-franc piece. Or they can be exchanged for things of a different category—a 5-franc piece can be exchanged for a loaf of bread.

The value of the 5-franc piece is not innate, as coins rarely contain metal to the value they represent. Instead the value of a 5-franc piece lay in it *not* being a 10-franc piece or a 1-franc piece. Neither is it a button, or a metal washer.

Saussure's quietly revolutionary insight was that the language with which we construct and reconstruct our notions of reality, does not refer to a neutral, objective external world. It really only refers to an *internal* world created and agreed upon by communities, reflecting their values and priorities. Communities choose how to divide up and represent the world through their use of language.

It was this insight that led to semiotics and then 20th century cultural theory, a discipline that looked at culture as something constructed, rather than being a spontaneous expression of our existence. As culture was the property of communities, cultural theorists believed the analysis of culture would reveal how communities understood the world, and what they valued.

An example of how values shift with language comes from panpsychic philosopher Rupert Sheldrake, who makes an interesting observation about the evolution of the English language. As any TEFL (teaching English as a foreign language) teacher knows, a common problem for non-native speakers is that English has multiple ways to express the same concept.

Following the Norman conquest, English developed from two roots, the Romance languages and the Germanic (and later, Norse languages that came with the Vikings) giving native English speakers competing ways to express themselves.

Sheldrake points out that, to this day, in both German and French, the majority of nouns have a gender. These apply to animals, inanimate objects, features of the landscape, even the planets. But when you combine two languages these genders will sometimes clash. For example, in German the moon is masculine, the Sun is feminine, whereas in French the genders are reversed. Sheldrake suggests that in order to simplify things, after the Norman conquests, objects took on the neutral gender: *she* and *he* increasingly became *it*. After a time, in England, only people and ships were left with a gender.

By removing gender, a mental map of the world is created for the English speaker in which anything that is not human is perhaps more likely to lack a personality or a story. I would suggest this makes the idea of mind in the non-human world more remote in the English language than in others. English is now the dominant international language of science and commerce, due to British colonialism and influence in the 18th and 19th centuries, followed by American culture dominating the 20th century. These are major factors in the success of English of course, but another element may be its extensive gender neutrality.

As northern Europe in particular accelerated towards industrialisation, a language that treated nature, and everything within nature as an object, not a gendered thing with a potential mind or life, was a better fit for a society that was increasingly treating nature as a profit-making resource.

The medieval *minds everywhere* concept is to an extent still being expressed daily by other European languages, where gendered nouns are widely used. In this sense, the gender neutrality of the English language represents, linguistically speaking, a break from animism.[63]

In the 1950s, the French philosopher Roland Barthes extended Saussure's observations of language, into a critique and decoding of mass culture. At this time, the mass media were generally assumed to be describing the world as they found it. It is partly thanks to the movement started by thinkers like Barthes, that large sectors of the population now assume the mass media rarely tells the truth, or is even conspiring to advance a specific political and economic agenda.

Barthes famously analysed the cover of one edition of *Paris Match*, in which an African boy in a beret is looking upward, saluting what is assumed to be an out-of-shot Tricolore. At a time of waning European influence over its colonies, Barthes addressed the subtext of this cover. The *myth* being promoted was that despite colonialism's shortcomings, French colonialism and militarism should be seen as a force for good. After all, here was an African boy, thousands of miles from France, loyally saluting the French flag. It is a myth that glosses over the ugly history of European colonialism. It is also a narrative the reader could subscribe to by paying the fee for having *Paris Match* delivered through the letterbox. All of this mythologising was expressed by a single photograph on a magazine cover.

To clarify, *myth* here does not mean fantastic tales from a long lost civilisation. Rather, a myth is a story within a culture that seeks to be accepted as a universal truth. In his 1957 book *Mythologies*, Barthes analysed the obviously selective realities of advertising and political campaign photos. He also tackled less apparent myths, such as the one around the French wine culture. Barthes argued that red wine, in particular, was presented by advertisers and politicians as the essence, almost the life-blood of French society. As a unifying symbol of Frenchness, he believed the mythology of red wine was papering over the cracks of inequality.

Barthes was a leading figure in a type of critical theory known as Structuralism, which treated culture as a series of structures in which messages are received. The Structuralists intended this approach to become a discipline, even a science of culture. However, because cultural theory encouraged the most radical re-examination of every aspect of accepted culture, Structuralism itself was (perhaps inevitably) then torn down in the late 1960s, to be replaced by Post-Structuralism, spearheaded by the brilliant, often impenetrable, works of French philosopher Jacques Derrida.[64]

As intellectuals moved from linguistic theory in the first half of the 20th century to cultural theory in the second half, they remained an almost exclusively left-wing club. Writers like Barthes, John Berger, Derrida and Julia Kristeva analysed how culture reinforces inherited power structures—the wealthy over the poor, the patriarchy over women, the coloniser over the colonised, etc.

However, although these theorists made it their mission to challenge the cultural narratives that determine our sense of who we are in the world, in common with the majority of Western thinkers, they rarely addressed another power structure that deserved cultural analysis—that of the human species over other animals, and over nature itself.

A choice of signs (words) is not just about how precise the author wishes to be in defining an idea or concept. These signs come with a pre-existing value, and sometimes an underlying narrative that becomes a myth, if that narrative is accepted as a description of reality. Returning to David Chalmer's *hard problem of consciousness*, it makes a value judgement about the process of finding an answer to the mind/body problem, because Chalmers tells us the problem is "hard".

A hard problem is presumably one that can only be figured out with a long slog of analysis, enquiry and examination, like a fiendish mathematical proof. By definition, it would not have a simple answer that a child would instinctively understand—such as there being an element of mind in all things. It suggests the answer will only be available to highly trained analytic minds, with PhDs from prestigious universities. For me, labelling the mind/body problem "hard" adds to the myth of human consciousness as something elusive and remote. I believe it is, in fact, a prosaic product of the everyday material world.

One potential myth about human existence is that we are exceptional as a result of an unusual level of interconnectedness in key areas of our brains. It has been suggested that a *step function* might represent the increasing number of neural connections as humans evolved from other animals, and that this might be the key feature which elevates our minds above theirs.[65] A step function in statistics is a relatively small change in input that leads to a dramatic change in output. Imagine a simple graph, with X and Y axes, onto which you draw a steadily climbing line. (Fig 8). At some point a threshold is reached, and the line goes sharply upwards, almost vertical, so the steady incline suddenly includes a step.

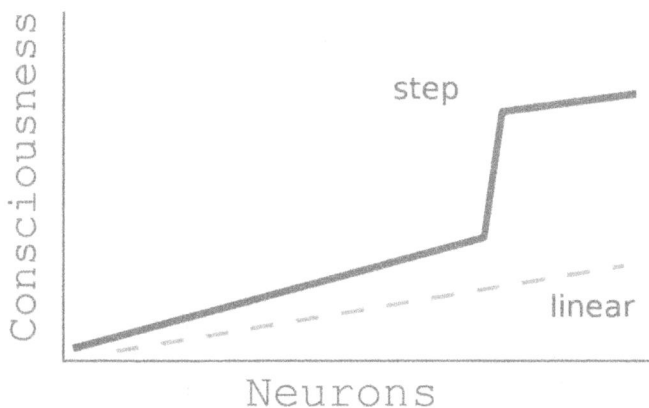

Fig 8. A step function. Here the input is the number of neurons, the output is the level of consciousness.

In terms of the human brain, the evolution of our frontal cortex supposedly caused a dramatic change in the inter-connectedness of our brains, which elevated our consciousness, and so accounts for the extent of human achievement.

Whether or not a count of neural connections should be represented with a step function is something for the neuroscientists to determine. However, we need to be aware of the potential for myth-making. Is the driver behind this idea good objective science? Or is it just another incarnation of the myth of human separateness that our culture has inherited from the monotheist religions?

Imagine that graph again. Intelligent consciousness is the output, but populate the graph with species, starting with the neurologically simpler animals on the left.

As the more "complex" species are added, the line of intelligent consciousness climbs steadily as we add fish, small birds and mammals, then clever birds like corvids and parrots, larger mammals like cats and dogs, maybe the octopuses. Near to us we place the primates, elephants, whales and dolphins, then larger apes. Throughout, the line continues to climb steadily. But when humans are added, there is the characteristic rise of the step function, as the line goes near vertical. (Fig 9).

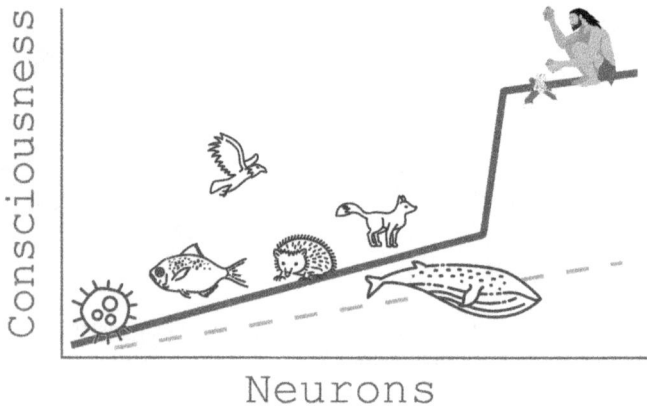

Fig 9. The step function as a myth.

Viewed like this, this step-up becomes a sudden leap up the cliff of intelligent consciousness. Throw in some swamps or forests to complete the background down below, and the picture being painted is one of us sitting on top of a cliff, up there with the most privileged and enlightened viewpoint in the animal kingdom. Here we apparently sit, watching over all the other species, who have been left behind, quietly grazing on the lower slopes of awareness.

I am deliberately exaggerating the idea of a step function in this context, because exaggeration helps to identify a cultural myth. In earlier chapters I argued that it is reasonable to see human consciousness as a subset of awareness in an essentially aware universe. A step function suggests there is a threshold only one species has crossed, creating a hierarchy of consciousness that makes widespread awareness into a remote possibility.

The myth of the innate superiority of the human brain is clearly one that Western culture has nurtured. There is a persistent idea that we only use 10% of our brains, and that the human brain has vast dormant potential, which could lead to wonders such as world peace, cures for all known diseases and interplanetary travel—if only we could tap into that unused capacity. In 2003 the neuroscientist Suzana Herculano-Houzel surveyed college students and found that more than half thought this was established fact. It is in fact a myth. In normal life we use 100% of our brains at some point.

Scientists are not immune to such myths, either. Early in her career Herculano-Houzel had difficulty finding an accurate figure for the number of neurons in the human brain. Asking around her fellow scientists, consensus settled on a figure of about 100 billion neurons. However, no one could really say how this number had been arrived at. For the first decade of the 21st century, 100 billion neurons was the widely accepted number among experts, although Herculano-Houzel found the scientific evidence for this figure to be patchy.

The established method for determining brain composition has been to section a brain into thin slices, then use a technique called stereology to make that two dimensional image more three dimensional, in order to estimate the density of neurons. This is unlike, for example, slicing a fruit cake to count how many raisins it contains, because although you can slice a cake to the thickness of a raisin, you cannot slice brain tissue to the thickness of a neuron. Analysing small samples with stereology therefore means there is a risk that an uneven distribution of neurons in the sample will skew the numbers.

Herculano-Houzel's answer was to create the "brain soup" method. By turning brains into a liquid, even a small drop should contain an evenly distributed number of neurons. Brain tissue was initially processed by dropping it into a blender. (Herculano-Houzel later developed a more subtle technique for that step!) Next, formaldehyde was added to protect the neurons, the liquid removed, and luminous dye was added to reveal the neurons for counting. Keeping with the culinary metaphors, Herculano-Houzel's method allowed for a more accurate assessment of the brain through *gazpacho*, as opposed to the long established technique of *carpaccio*. She began by analysing rodent brains, and made some interesting discoveries. For example, rodent species with smaller brains have an increased neuron density. Also, while the raccoon may be a rodent, its neuron density is closer to a primate.

What was really ground-breaking however was her analysis of primate brains, which turned out to have a consistent density of neurons, whatever the brain size. Using the brain soup method, she also had the opportunity to analyse four human brains. The largest was found to have 91 billion neurons, while the other three had about 86 billion. Those counts are considerably below the 100 billion quoted by many scientists prior to her research. The reason this difference is significant is that purely in terms of a neuron count, there is no *step-up* from the other primates to us. With 86 to 91 billion, the human brain has a density of neurons consistent with other primate brains. Conversely, the until-then accepted figure of 100 billion neurons would mean we had been blessed with a brain of unusually high density. That figure is now accepted as incorrect.

That this number has been wrong, well into the 21st century, shows our species is not always able to correct for its own biases. When scientists relied on the less accurate method of stereology for a neuron count, might this inaccuracy have created a tendency to revise estimates upwards? This would have matched the widely held belief that the human brain is in some ways the exception to the rules of nature, and confirmed what we already believed about ourselves.

The 100 billion neurons figure has been quoted in academic works without citation, which further fixed its status as basic scientific fact. With a more accurate count however, our brain reveals itself to be just another primate brain. As Herculano-Houzel writes in the prologue to her book *The Human Advantage*:

> Comparing the human brain to the brain of dozens of other animal species large and small has been a most humbling experience, one that reminds me that there is no reason to suppose that we humans have been singled out in our evolutionary history or 'chosen' in any way.[66]

In discussions of consciousness, another frequently asked question is, *what is the one thing that distinguishes humans from other animals?*

Clearly there are significant differences between human and animal minds, but also a growing understanding of the many similarities. Yet in a strange way, the greater the number of differences claimed, the less special we become, as the battle to keep our elevated status is fought on many fronts. Any trait claimed to be exclusively human, like self-recognition, moral behaviour, abstraction etc. faces a fresh challenge to its human-only status.

However, because our way of living is so different, other animal abilities do not always map onto ours when we seek to make comparisons. This leads to debates about whether, for example, a dog recognising its own scent is the equivalent of a human recognising themselves in a mirror. In examining these differences, attention is then drawn to the similarities, so the picture, and our separation from other species, becomes less clear-cut.

In order to preserve the idea that we are superior beings, there is a tendency to claim there is a *singular* key that unlocks the door to our world of experience. It is as if behind this door is the bright piercing light of human-only consciousness. There has been an underlying narrative in Western culture that the essence of a human being is not just different from the essence of a rat, dog or chimpanzee, which we can expect to differ from each other, but also that this essence is somehow *transformationally* different. It is the secular equivalent of the soul, which was the pre-Enlightenment essence of a human being when religion set the moral and intellectual agendas. One reason discussion often focuses on "the one thing" is because in secular society, it is the *one* thing—the step-up in connections in the PFC for example, or a conceptual leap made by a distant primate ancestor—that serves as a proxy for the soul. It is the *one* thing that supposedly makes us special, not the many things.

The choice of what to identify as "the one thing", be it self-awareness, abstract thought, theory of mind or meta-knowledge has of course changed, as one by one, animal research has shown such divisions between us and other species are mostly false.

Adult chimpanzees are in most respects considered to be intellectually at the level of 2-3 year old human children. But some chimp abilities actually outstrip ours. A few years ago a chimpanzee in Japan, named Ayumu, demonstrated what in humans might be considered genius level ability in a memory test of numbers.

When shown numbers from 1 to 9 randomly distributed on a screen, Ayumu can tap out their sequence from low to high. Even with practice, humans trying to match Ayumu's success with nine numbers reach their limit at around five numbers. Ayumu is also much, much faster. He manages several numbers in just over a fifth of a second, compared to the several seconds taken by humans. Had the results been the other way round, and humans were found to have this ability at a superior level to other primates, it is inevitable some would have claimed this as another example of our exceptionalism.

I started this chapter with Arthur C Clarke's *2001: A Space Odyssey*, and said whether he actually believed the dawn of human civilisation included visits from aliens was not important. Yet millions of people across the world believe that early human development was accelerated by extra-terrestrials. I believe this is another consequence of our tendency to mythologise human consciousness.

What is known as the "ancient astronaut" hypothesis became popular in the 1960s and 70s thanks to books like *One Hundred Thousand Years of Man's Unknown History* by former sci-fi writer Robert Charroux, and *Chariots of the Gods* by Erich von Däniken. These books belong in the category of pseudo-archaeology, because the supporting evidence is sketchy at best. While true archaeology is built on the steady accumulation of evidence over many decades, pseudo-archaeology works through a cherry-picking of artefacts and evidence to support a specific narrative.

Despite this, the ancient astronaut hypothesis continues today, fed by content online and on cable TV. There is an undeniable element of racism here too—why claim the Egyptians or Mayans needed help from aliens to build their civilisations, but the Ancient Greeks and Romans did not? Pseudo-archaeology is worth mentioning here, however, because I believe another important form of bias is involved.

The ancient astronaut hypothesis may have been given some unintentional impetus by the legendary scientist and writer, Carl Sagan. Sagan was one of the founders of the non-profit SETI organisation, which searches for signs of extraterrestrial life. In 1966, he co-wrote a book called

Intelligent Life in the Universe, with the Soviet astrophysicist, Iosif Shklovskii. The book covered a wide range of issues in cosmology, including possible forms of life on other planets in our solar system and beyond, and the "panspermia" theory, the idea that life came to Earth from another part of the cosmos.

In the spirit of considering every possibility, one of the book's 35 chapters imagined a race of advanced beings, beings capable of interstellar travel, who might keep a careful watch on the development of lifeforms such as ours, throughout the galaxy. As humans developed, the aliens would increase the frequency of their visits to every few thousand years.

Sagan then wondered how such visits might be received, and drew a parallel with the accounts of first contact between Europeans and indigenous people, such as the first meeting of French sailors with the Tlingit people in Canada, in 1786.

In the late 1800s, an anthropologist had documented the Tlingit's oral version of this encounter. Nearly a hundred years on, the Tlingit themselves described the sailing ships as great black birds with white wings. Other than this, the details from their oral account largely tied up with the written account by the French sailors. Consequently, Sagan's idea was, if the interval of a few thousand years was right, previous visits from extraterrestrials might show up in the histories of ancient civilisations as visits from strange, even god-like creatures, bearing gifts of wisdom and technology.

Looking at the historical records, Sagan gave the most attention to the Sumerians, an early civilisation, starting around 4000 BC, whose language had no known roots. The Sumerians played an important role in the development of early civilisation, as they were the first step in a lineage that led to the Mesopotamians, then the Assyrians, then the Babylonians by around 2000 BC.

According to Greek scholars, Sumerian history described several visits from a two-legged fish-man called Oannes, who came out of the Persian gulf and instructed the Sumerians in science, art, law, geometry, agriculture and building techniques, giving them all the elements needed to build their civilisation. The Sumerians also had the legend of the Anunaki, giant god-like beings who came down from the stars, wore horned hats, and were covered in a substance called Melam, which accentuated their powers. Sumerian writing is on clay cylinders, and some claimed the drawings on these cylinders show the Earth orbiting the Sun, thousands of years before Copernicus. Even after bringing all these points together, Sagan raised the idea of any civilisation having documented alien visits with a significant level of caution, even in the case of the Sumerians.

However, best-selling books like *Chariots of the Gods* in 1968 claimed that there *was* strong evidence for extra-terrestrial intervention in human development. Authors cited Sumerian culture, sections of the Bible and *The Book of Enoch*, which at one point describes 200 angels descending from heaven and interbreeding with humans. These writers also claimed that great human achievements such as the Easter Island heads, Egyptian pyramids and Stonehenge, were beyond human technology at the time and must have been built with alien assistance.

Sagan was perhaps irked that alien intervention in early human history was being claimed as near certainty on the thinnest of evidence, and in 1979 called such writers "uncritical".[67]

Today, there remains precious little evidence for ancient astronauts. The hypothesis is only credible if we focus on certain stories and artefacts from early history, give them a specific interpretation and ignore more credible, more prosaic explanations. For example, archaeologists have since demonstrated how the stones that built Stonehenge were likely transported by river, then dragged to the site with only ropes and wood.

The great pyramids of Egypt may have been built using ramps of sand, with extra sand continuously added, forming a spiral, as the pyramid grew higher. Although we now credit Copernicus with single-handedly correcting our understanding of the solar system, the Greek astronomer Aristarchus of Samos knew from his calculations that the Earth orbited the Sun.

Whether or not they documented it, there could have been other civilisations whose dominant religion was not threatened by an understanding that the Earth was not the centre of the universe, as the Christian church clearly was in Copernicus' time. Medieval Arab scholars understood this, as evidenced by manuscripts found in Timbuktu, Mali.

Pseudo-archaeology is for many people more real than decades of accumulated archaeological study and carefully sifted evidence. Humans have a propensity to believe explanations that come with imaginative stories, especially those in which we can picture ourselves having a more interesting life than the one we live now. By the late 20th century, the ancient astronaut hypothesis was a core belief for several cults, including the southern Californian group Heaven's Gate. Sadly, these stories were all too real for the cult members. Of their 41 members, 39 participated in a co-ordinated mass suicide over a period of three days in a house in a San Diego suburb in 1997. They believed, through suicide, they would then be taken on-board an alien spacecraft, while comet Hale-Bopp passed by.

The lack of evidence makes it easy to dismiss ancient astronauts as the territory of UFO cults, conspiracy theorists and those with a strong need to believe in religious stories, no matter how fantastic. But I have not described it here as an example of the eccentricity of human belief. Instead I am suggesting its primary driver is in fact a *mainstream* force in Western culture. In a culture that has held us up as the exception to the rest of creation, it is easier to believe such exceptionalism cannot come from the ordinary atoms, soil, rocks, or simple organisms of the everyday natural world that we see around us. Ancient astronaut believers are as likely to rubbish scientific evolution as they are Christian creationism—because Darwinian evolution firmly places the roots of our clever minds here on Earth, not somewhere *out there* in the cosmos.

If humans are so remarkable that we could *only* be the product of alien intervention, does it not follow that an alien species far more advanced than ours (after all, they are capable of interstellar travel) could not have developed organically on *their* planet without outside intervention? Either their planet is a far more remarkable place than Earth, as it is capable of creating advanced, galaxy-roaming species from basic matter —or they too must have had alien benefactors to have developed their wondrous technology. Like the homunculus living in our heads, that leads to a chain of causes and effects that can never be resolved.

Despite these logical problems, having been brought up to consider ourselves in some way removed from the rest of nature, for many it seems feasible that an alien civilisation might have dropped in on our ancestors. Thanks to the myth of human exceptionalism, it then seems feasible that they gave our species, and our species alone, an evolutionary leg-up.

How did humans change from clever groups of hunter-gatherers to nation states containing a myriad of secular and religious groups, with global culture at the centre of everything we do?

Our human ancestor branched off from the other apes around 5 million years ago and brain sizes increased as we became bipedal. One theory for this increase is that walking on two legs needs less energy than walking on four, making more calories available to the brain. In addition, humans can lose heat more easily than other animals due to a large number of sweat glands. This allowed for persistence hunting, where large, fast animals can be steadily pursued, perhaps for a day or more, until they collapse from heat exhaustion, or be forced into marshes or other natural traps for an easier kill. The additional animal protein from these hunts again rewards brain size, increasing intelligence.

For over a million years, early humans were making a limited range of tools, such as hand axes and points. Around 300,000 years ago, Neanderthals began making Levallois points, which take great planning and skill to successfully execute. The stone must be worked carefully, while keeping in mind a clear idea of how to reach the finished article. The point itself is then formed with a single final strike, a skill that few present day humans have developed.

In the 1970s, Robin Dunbar proposed in *The Social Brain Hypothesis* that primate brains grew large because as group sizes increase, a larger and more complex brain is needed to keep track of increasingly complex relationships. As we became more social, having a theory of mind of other animals would have given our hunter-gatherer ancestors an advantage over other species, as they anticipated the behaviour of their prey. Some believe crediting other species with minds is therefore an instinctive mistake, something we are inclined to do without evidence, (the "folk wisdom" mentioned in the previous chapter) and regard this as a vestige of the early human mind, part of our primitive toolkit.

But early human neurological hardware was probably not all that different to ours, with brain sizes similar to modern humans (Homo Sapiens). Neanderthals, in fact, had larger brains than modern humans, and even Homo Erectus and Homo Habilis had Broca's brain areas (for the production of speech) and similar vocal tracts to modern humans. Archaeologists estimate that group sizes for early humans species were slightly lower than those of the modern humans that followed.

However, despite early humans' technical skills and sociability, early human technology stayed fairly static until some 300,000 years ago, and was mostly restricted to stone points and hand axes.

The earliest modern human skeletons date from 200,000 years ago. Their bones and brain size were essentially the same as ours today. They made tools and were well adapted to their environment. But human symbolic behaviour only made an appearance some 100,000 years ago, when, for example, pieces of ochre and ostrich shells were being marked with geometric patterns.

The archaeological record shows that there was then an explosion of human tools, art, music, decorative and religious artefacts worldwide, some 60,000 years ago. This came with the growth of language, as modern humans left Africa, moved into the middle East, then Europe. Symbolic and religious thinking accelerated. There were ritual burials, such as an adult and two children buried 28,000 years ago in Russia, wearing thousands of beads made from mammoth ivory. This is a significant amount of time and effort to invest in something without practical value, as each bead probably required an hour's worth of work. Leaving these buried with a corpse suggests that for these people, great importance was given to the afterlife, to a world they could not see.

Archaeologist Steven Mithen's book, *The Prehistory of the Mind* has an interesting take on how humans developed our culture dependent mind 60,000 years ago.

Mithen argues those early human species had several modules of intelligence devoted to specific domains: the technical; the social; the natural history (knowledge of other species' behaviour). This leads to a generalised multi-purpose intelligence, learning by association, or trial and error. As humans developed, intelligences might have become more specialised, such as the high level technical intelligence required to make Levallois points. But for most of early human development, these intelligences were not significantly linked together. What he believes allowed us to shape the world is that these multiple intelligences *leaked* across domains into one another, allowing for greater insight and creativity.

For example, Mithen points out that of the animal bones that came from early human hunts, precious few have been worked. Our ancestors rarely attempted to turn the bone or ivory gained through a hunt, into tools for hunting or other uses. Perhaps they did not consider bone as a potential tool making material, because the animal and its carcass were part of the social history domain, not the technical domain, which was only familiar with stone as the hard part of a tool.

Mithen also observes that chimpanzee groups can have complicated dynamics and politics, but do not use the material world to create or reinforce their social order, as humans were doing some 60,000 years ago. This he regards as a missed opportunity. Chimp grooming increases significantly when the group is under pressure, and chimp society is built on close personal interactions, where time spent grooming is an important investment. The problem is, a chimp can only groom one individual at a time.

For humans, language or symbolism enable one individual to develop and maintain relations with many group members at once. Chimps don't use material status symbols to re-enforce the social order, which suggests to Mithen that these different intelligences are not interacting with one another.

In modern humans, symbolic behaviours, and even religious belief itself, could be a by-product of more extensive tool making. Returning to the point in human development when symbolic behaviour began, 100,000–60,000 years ago, when our modern human ancestors left Africa, there is a question of how religious thinking came about.

We might assume that early religious beliefs preceded the making of early religious artefacts—that the artefacts were the expression of already long-held belief systems. However, it has also been suggested that as the human mind developed a greater ability to imagine tools and other *objects* that did not yet exist, it was matched by an ability to also imagine spirits, gods, and other powerful beings and realms.[68] These immaterial worlds and beings, these inventions of the mind, then became real to our ancestors through the considerable power of the modern human's expanding imagination.

Dragging the timeline back up-to the present, we have good reason to claim we are the most intelligent species, as our abstract imaginations have given us the evolutionary advantage, resulting in a population explosion, which is certainly one measure of evolutionary success.

But there are also reasons to question whether our imagined and abstract concepts represent evolutionary superiority. Abstraction is humanity's most powerful tool, a piece of mental machinery that leads to the most profound transformation of everything around us. As with all tools it can also be destructive when misdirected. In the 20th century, the technology for killing our fellow human beings gave us the capacity to butcher tens of millions through war. The dictatorships that plagued the 20th century always managed to find abstract ideological justifications for killing thousands, or even millions, of their own people, even in peacetime.

Thanks to our abstracting morality into competing political ideologies, the world came perilously close to all-out nuclear war during the Cuban missile crisis in 1962. The writer Gore Vidal was related to the Kennedys and dined at the White House. In his autobiography *Palimpsest*, he rather presents John F. Kennedy as fully prepared to use his nuclear arsenal.

On the Soviet side, one commander, Vasily Arkhipov, single-handedly prevented nuclear war. When his submarine lost contact with the fleet, the senior officers on board thought full-scale war may have broken out, and wanted to launch nuclear weapons at the USA. Mercifully, the launch protocol required the captain, political officer, and the commodore of the flotilla to agree. Arkhipov, as commodore of the flotilla refused to launch. Had Arkhipov been more vulnerable to fear and group-think, an outbreak of paranoia on one submarine could easily have led to full-scale nuclear war.

In 1962, a contained nuclear exchange was perhaps still possible, but warhead numbers increased rapidly in the 1960s and the doctrine of Mutually Assured Destruction (MAD) became the norm until the Soviet Union collapsed. At the start of the 21st century we might have been able to excuse the lowest points in human history as tragic bumps on the road, as humanity worked towards a more ideal state of being—what the US constitution calls "a more perfect union"—thanks to technological and economic development. There was a growing sense at the end of the 20th century that the ability to communicate instantly with anyone on the planet would almost inevitably lead to mutual understanding and more equitable societies across the globe.

Decades on, however, millions die from hunger and lack of adequate health care every year, and we are still failing to address the major threat to all of humankind's future. The 2006 Stern Review on *The Economics of Climate Change*, prepared for the UK government by the economist Nicholas Stern, estimated that the world could be spending up to 20% of its wealth dealing with anthropogenic climate change. Progress on mitigating climate change currently lags far behind where the majority scientific opinion says we should be, and it is not the only danger to the ecosystem we depend upon. Earth Overshoot Day, the day in the year that marks the consumption of one year's worth of the planet's natural resources, is being reached earlier every year.

Gross domestic product is a measure that takes no account of poverty, income inequality, or the long-term health of the environment. Yet, GDP is now *the most* influential of all human abstractions. For decades, short-term measures of GDP have taken precedence over dealing with the actual health of the only planet capable of sustaining human life in the known universe. There is an inherent stupidity in our species' inability to face up to the environmental problems we have created, which now puts our own species at risk. If we really are on top of the cliff of intelligent consciousness, as the myth would have us believe, it sometimes feels as if we are throwing matches down onto the dry tinder below, with the idle curiosity of wanting to see what might happen next.

Increased communication has increased social division as much as it has understanding. Social media algorithms reinforce our existing world views, and actively promote content that causes outrage. This emotional manipulation maximises time spent on their platforms, further strengthening cultural and ideological silos. We have moved from consumer choice about goods and services, to a far, far more dangerous form of choice—a consumer choice about what we are willing to accept as basic reality.

Many conspiracy theories are baseless, but do at least have a comprehensible driver behind them. The politicisation of public health measures, including mask mandates in the 2020 COVID-19 pandemic largely served as an excuse that gave people permission not to do something they disliked. The growth in QAnon and online conspiracies around the 2020 US presidential election could have brought the end of over 300 years of American democracy, because millions of people were unable to accept that one of the most unpopular presidents in US history had actually lost an election.

More bizarre beliefs show how the rational mind can be used to justify what are really only statements of faith. There is very little advantage for anyone in being a flat-Earther. It is not the route to wealth, political power or fame. Its only purpose seems to be to justify the believers' paranoia that all authorities routinely lie to the populace about the basic nature of reality. The only pay-off is a sense of kinship with others who share their belief.

But the alternate realities created by social media do not just mark an increased cynicism about governments, the law, or faith in democracy. These alternate realities cost lives. For example, a 2018 UN report found that deliberate misinformation on social media was a major factor in systemic violence against Rohingya Muslims in Myanmar.

In the end, as a species we are highly susceptible to the hypnotising effect of our own abstract creations. The extent to which we put our faith in GDP or tribal politics, often based on highly selective versions of reality, may yet be an evolutionary experiment that backfires on us spectacularly. This will certainly be the case if we are unable to quickly find a consensus for dealing with our profoundly damaged planet.

The abstract ideas that underpin human culture can represent an ideal to strive for, or they can stand as a symbol of something that is temporarily absent. Yet as we have found in the 21st century in particular, it can also be used to manipulate the emotions, to con, cheat, lie and destabilise democracy with false realities, which become fixed as truths in the minds of millions.

Viewed in this context, the chimpanzees' failure to express their social order with material symbols may be a major difference between us and our nearest relative, but is hardly a sign of human superiority. We need a reality check on the value of early 21st century human culture. Where is it leading us? More importantly, where is it misleading us?

8

The *why* of existence

Having argued through this book that we may live in an aware and behavioural universe, and that intelligent awareness may be an inseparable attribute of all matter, one question looms large: *Why?* Why would this be the case? What ultimate purpose would universal intelligence and awareness actually serve?

As stated previously, this is a book about the *how* rather than the *why*, and I have no insights into a purpose for life, human or otherwise. Intelligent awareness could exist with only a *how*. It does not necessarily need a *why*. But as the main purpose of organised religion is to answer the question of why we are here, and as mind in all things is a strong theme in the mystical traditions of all religions, the question is worth looking into.

One form of panpsychic philosophy is pantheism, a type of religious mysticism that entails belief in a deity that is present in all things. Some will claim that an aspect of mind in all things requires nature to be the work of an intelligent designer. Believers argue that God gives us our moral purpose, which is the *why*, and that a form of intelligent design therefore gives us the *how*, to complete the other side of the equation.

The intelligent design (teleological) argument came to the fore as the Enlightenment heralded the scientific age, and science began to decode chains of cause and effect in the natural world. Nature was treated as a set of interconnected and moving parts, operating under finely balanced rules and conditions. Rather than working to undermine God, many scientists believed they were doing God's work by seeking to understand and describe the divinely inspired mechanisms that governed the natural world. They used the teleological argument to explain God's power in nature. This says that as nature has a self-regulating harmony, where species are ideally suited to their circumstances, and because there are variations on a theme, such as eyes being of similar design across most species, nature must have been guided to this end by a divine being with the intention and ability to design—an intelligent creator God.

One response to the intelligent creator argument is that if the designed elements of nature must by necessity be the product of an intelligent creator, why does the creator not also need a creator? If the creator needs no creator, isn't that engaging with cause and effect just long enough to reach the conclusion the believer wants? God is then being defined into existence as the necessary creator of all things, who needs no creator. Intelligent design arguments largely reinforce the faith of those who already have faith. Perhaps this is as it should be. Religious faith is not meant to be an exercise in logic or reason, and I do not say that to denigrate religion. Far from it. Faith is concerned with the powerful, and frequently empowering, need to believe in something greater than ourselves. The purpose of faith is hope, not reason, and hope is a very potent force in all our individual and collective destinies.

However, another major driver behind intelligent design is the idea that the human species is the *exception* in nature, that we are blessed with something other-worldly, which then requires something not bound to this Earth to explain our existence.

In secular terms, Darwinian evolution through natural selection is as close as science gets to addressing the *why* of human existence. Early on, Darwin was influenced by the work of theologian William Paley, who gave a well-structured teleological argument in his 1802 book *Natural Theology*. It was Paley who said that if we found a watch in the middle of nowhere, we could infer by its complexity that it had been created by a skilled watchmaker. By analogy, the wonder and complexity of nature should be taken as indicating an intelligent creator must have been at work.

Before his departure on HMS Beagle, the young Darwin had considered a career in the clergy. However, after five years away, he found his observations to be incompatible with a divine creator, and said this of the teleological argument in his autobiography, written in 1876:

> The old argument from design in nature, as given by Paley, which formerly seemed to me so conclusive, fails, now that the law of natural selection has been discovered.

Darwin was not the first to advocate an evolutionary principle. It was an idea that had been around for many decades. He had collaborated with fellow naturalist Alfred Russel Wallace on a paper for the Linnean Society that outlined the principles of natural selection, after both men had independently reached the same conclusion about how evolution could work in practice. But it was the detailed evidence that Darwin had gathered on the voyage of the Beagle, and twenty years of his own experiments breeding pigeons and collating evidence from other naturalists, that documented evolution in action and the mechanism of natural selection. Because Darwin possessed the strongest body of evidence, it was he, not Wallace, who wrote *On the Origin of Species*, and became known as the man who had "killed God".

However, if Wallace had published first, he may have avoided some of the accusations of godlessness that Darwin received. Apart from Darwin's additional evidence, the two scientists differed in another important respect—Wallace believed that modern human development had been the exception to evolutionary forces. Although an important scientist, Wallace, unlike Darwin, believed in the myth of human exceptionalism.

Evolution is such a part of established science that some misconceptions have grown up around it. For example, evolution is not driven purely by random chance. Predicting which number will land face up when we throw a dice is considered chance, even though the dice is governed by the repeatable laws of classical physics. Similarly, evolution is said to be random because there are too many variables, too many forces to account for, meaning the outcome should in practical terms be attributed to chance. However while the cause of a genetic mutation that leads to an animal having different characteristics from its parent may be considered chance (in practice) the evolutionary process itself—and the ability to pass that adaptation on to later generations—is not chance. It is because the organism that is best suited to its environment is the one most likely to survive.

Another misconception is that Darwin bears some responsibility for modern-day greed, selfishness and amorality. Darwin has wrongly been associated with either right-wing, or godless ideologues, who claim the unsentimental mechanism of "survival of the fittest" is somehow a model for how society should organise itself, taking it far beyond its intended purpose. This is the antithesis of a principle to which I subscribe—variously credited to Dostoevsky, Gandhi, the novelist Pearl Buck and US vice president Hubert Humphrey —that the measure of a civilised society is how well it treats its most vulnerable members. Reading Darwin, there is no suggestion that natural selection was his recommendation for building our collective futures.

It is worth looking at the question of purpose in evolution, by contrasting Darwinian evolution with an evolutionary theory proposed by the French zoologist Jean-Baptiste Lamarck, half a century before *On the Origin of Species* was published.

Although their theories differed greatly, Lamarck's work had helped prepare the ground for Darwinian evolution. Lamarck saw commonality between species and argued that this was because species had evolved, rather than having had a fixed state since the beginning of time, as most of his contemporaries believed. He was certainly ahead of his time there.

Still, Lamarck's theory was based on several wrong assumptions. For one thing, he believed that species would not become extinct by natural means.[69] Strange species that were being found as fossils were to Lamarck likely still in existence, out there somewhere, waiting to be discovered in the Earth's unexplored oceans and forests.

What Lamarck is famous (and infamous) for was his belief that the evolutionary process was driven by habits of use and disuse during an organism's lifetime. This means an organism's decision making and even intentions are the primary drivers of evolutionary change. Lamarck argued that giraffes might be descended from antelopes, who in the process of constantly reaching for the higher branches, would stretch their neck muscles, and then pass this enhanced reaching ability onto their offspring, which over several generations would lead to the giraffe. He also believed moles had weak vision because their subterranean ancestors had not used their eyes frequently enough.

The main problem with Lamarck's theory is that it makes evolution a fast-acting process. It suggests significant changes could occur in a few generations, which does not fit with the archaeological evidence for a slow progression of change in the majority of species. Lamarckian evolution sounds even more implausible in the context of another of his famous examples, the idea that due to the physical nature of their father's occupation, the sons of blacksmiths will be born with a tendency to have strong brawny arms! To believe this today would require us to ignore everything we know about genetic inheritance, or the effects of nutrition and exercise on a child's physical development.

However, Darwinian evolution by natural selection was by no means a complete theory with no contraindications, something Darwin himself was aware of. Although he turned evolution from an idea into a credible scientific theory by the weight of his evidence, when he died he had little evidence for how variations could be passed from generation to generation, which we now attribute to genes. Darwin theorised about gemmules, rather like small seeds, shed from every cell like grass seed in a field, with the potential to seed other cells, and which collected in the reproductive organs ready to be passed onto the next generation.[70] He was clearly off-target here, because none of those characteristics match what we now know about genes. On the other hand, his concept of *gemmules* shared some characteristics with genes, in that they carried large amounts of information, would often remain dormant, and were to be found in most living things in vast numbers.

Natural selection is often thought of today as a wholly unfeeling and impersonal process, without intention or plan, which happens regardless of an organism's intentions.

But that is our contemporary understanding, and on this point Darwin's view was perhaps not entirely Lamarck's polar opposite. In his later work Darwin did believe that events in a parent's lifetime could, in certain circumstances, influence the health of the offspring. This is known as the inheritance of acquired characteristics. In his book *The Variation of Animals and Plants under Domestication* Darwin wrote this in 1875:

> There can be no doubt that the evil effects of the long-continued exposure of the parent to injurious conditions are sometimes transmitted to the offspring.

Random change as the sole driver of evolution has become increasingly open to challenge, and like the intelligent design argument, this question comes with some philosophical baggage. Given that Darwin had far more evidence available to him than Lamarck, I suggest the most important difference between Lamarck and Darwin is perhaps not the different evolutionary mechanisms they described, but the personal philosophies that led them to their conclusions. Both clearly believed it was possible to inherit acquired characteristics. But in Lamarck's work *Philosophie Zoologique*, he regards evolution as a process of organisms working towards a form of divinely inspired *perfection*. Further, he regards humankind, as the most complex species, as the closest to this ideal.

Unlike Darwin, he found evolution to be compatible with a divine creator. However, unlike most of his contemporaries, who believed God had fixed the species at the moment of creation, Lamarck reasoned that a divine creator was more likely to have set in train a process of evolution, giving his own teleological answer to species diversity. The reason he believed species would not naturally become extinct was that extinction would suggest nature, and by extension, God, was capable of making mistakes. This reaching for perfection makes Lamarckian evolution a better fit for anyone who believes evolution, and our lives, must ultimately hold some spiritual purpose. Lamarckian evolution is very much open to the possibility of creation being God's work, whereas Darwinism keeps that door firmly closed.

However, the impression that Darwin believed only random chance drove natural selection comes in part from opposing camps of Lamarckians and Darwinians in the early 20th century. Supporters of Lamarck, who believed there was a moral purpose to life, cited randomness in Darwinian evolution as the major failing of the theory. On the other side of the fence, some Darwinians, for their own reasons, promoted the randomness in natural selection as the sole driver of change, in order to clearly and definitively exclude God, or any other form of divine intent, from human development.

In fact, inheritance of acquired characteristics was still very much on the table in the early 20th century. Even legendary scientists like Ivan Pavlov believed it was possible.

In 1943, in order to test whether an organism's genes could adapt as a direct response to their circumstances, the biologists Salvador Luria and Max Delbrück tested what happened to successive generations of bacteria that had been infected with a virus, compared to a control group. Through the experiment, they were expecting to demonstrate that inheritance of acquired characteristics was a reality. Yet their results indicated the opposite, as the level of variation in later generations was not significantly higher in the bacteria that had been exposed to the virus than those that had not. Bacteria with greater exposure were not passing on any increased resistance to the next generation, so Luria and Delbrück concluded that genetic mutations were therefore essentially random, and not adaptive. It is a measure of how this changed the consensus view in the mid 20th century, that in 1969 their work won them the Nobel prize for medicine.

However, the question of randomness vs possible intention in evolution has not ended there. Genes can of course be active or inactive, and genetic code alone does not determine what an organism becomes—you may have the gene for green eyes but not have green eyes yourself, so genetic code is not the only determinant of how an individual turns out. The first ever cloned cat came from a calico mother, who gave birth to a tiger tabby kitten, even though the cloning processes ensured mother and kitten were genetically identical. DNA is not an inflexible set of instructions. It is more like a recipe, in which many of the ingredients that make up the final dish are optional. Gene activity can also change without the genetic code itself changing, the study of which is epigenetics. This sometimes allows an organism's reaction to its environment to be passed to the next generation.

It is known that the children of parents who have experienced profound stress and trauma, such as holocaust survivors, have an increased risk of stress disorders themselves. In humans, it is hard to identify how much of this is due to upbringing, and the child's knowledge of their parents' trauma, and how much could be due to epigenetic changes.

In animals, studies suggest that upbringing may not be the only factor. At least one recent study in mice found that by forcing the parents to experience acutely stressful conditions, their body's stress response during pregnancy affected DNA methylation, which in turn increased sensitivity to stress in their offspring.[71] DNA methylation can affect which genes are active and which remain suppressed. Methylation, however, is not an example of Lamarckian adaptation. It is certainly not an example of the evolutionary process moving lifeforms closer to perfection, because its effects are generally negative. For example, doctors sometimes recommend pregnant women take folic acid to reduce the risk of DNA methylation.

Intriguingly, in some specific circumstances, scientists have found adaptive changes to the DNA itself. In 1988, researchers were working with Escherichia coli bacteria that were genetically lactose intolerant. The E. coli were left in a Petri dish where lactose was the only available food source, but they did not all die of starvation as expected. Some bacteria of later generations survived, because they had actually *fixed* the gene that caused lactose intolerance. A later study found a similar talent for gene repair in E. coli that should have been unable to thrive in tryptophan, because some of these bacteria repaired the faulty gene when tryptophan was their only food source. It seems that in certain circumstances, bacteria can repair a faulty gene when needed. This means that sometimes species change does not have to wait for a random mutation to appear.

Some view these experiments as evidence for a guiding mind behind nature, and a divine purpose to our individual and collective destinies. But bear in mind that this is only evidence of a faulty gene being repaired, not of a new adaptation happening within a single generation due to the parents' life choices, as Lamarck believed to be possible. Also, it's worth remembering that bacteria and viruses are some of the fastest evolving forms in nature—it is their main strategy for survival.

A philosophical view of mind in all things does not mean that there is then necessarily a divine purpose to life. I see a particular problem with the idea of mind in matter being misused in some New Age beliefs. You can think that mind, matter and energy are all the same stuff without believing your individual mind is therefore capable of bending the world to do your bidding, or that you can manifest the perfect partner, the perfect job, or parking space into your future. Our intentions, our beliefs and our conscious minds are only a small part of what makes us who we are.

As much as we may shape the world around us, the external world also has the power to shape us. New Age thinking frequently uses language that is on the surface empowering. It tells us we are in full control of our destinies. It puts us at the centre of the universe where the individual fate is the primary concern, as if the world we experience is being constantly built, torn down and rebuilt for us and us alone. In this sense, New Age ideas are usually idealist in nature.

With epigenetics affecting biology and bacteria repairing their faulty genes, some claim this is evidence for full-blown Lamarckism. Further, they claim we can change our biology by force of will, activating genes to fight disease using our will-power.

There is no doubt the mind has powerful influence over the body's ability to heal itself, as studies using placebos repeatedly demonstrate. However, where this philosophical approach to health becomes problematic is in relation to life-threatening disease, in particular cancer.

Cancer is a dreadful illness, and conventional treatments can be harsh and unpleasant to endure, so who would not choose a less painful and distressing alternative if it was feasible? Many people have claimed the power of thought prevents and even cures cancer, and some of them may themselves have survived cancer against unfavourable odds. Before anyone puts their faith in these claims, or imagines that a mind-in-matter view of our universe lends some weight to them, there is a basic statistical point to consider.

Worldwide there are some 14 million new cancer diagnoses annually. Averaged over all types of cancer, the odds of surviving are roughly 50/50. But imagine if the odds were much worse than this. Imagine those 14 million people annually are given a poor prognosis by their doctors, say a one in a hundred chance of survival. That would still mean 1% annually should survive, which adds up to some 1.4 million people over 10 years. Over a decade, 1.4 million people would seemingly have done the near impossible, and beaten cancer.

For every person who claims their thoughts have defeated cancer there will be many more making no such claims, and sadly millions more who will not live through the illness to make any claims either way. To make a rather obvious point, only those fortunate enough to survive cancer can then write a book, make a video, or charge for a ticket to a seminar where they tell a paying audience how they did it. Given the millions of people who are diagnosed annually, it is inevitable that thousands of people, at the very least, will survive in apparently unusual circumstances. Some of them, with the necessary self-belief and communication skills, may become New Age gurus.

But our intentions, our beliefs and our conscious minds only give us so much influence over all the energy and matter around us. In case anyone feels this book could be setting up an aspect of *mind in all things* as yet another New Age idea that claims to transform your life, it should be said that even if there were clear evidence for mind throughout the universe, taking this position will not necessarily add to your happiness! Be warned, this philosophical point of view may even make you *unhappy.* In general, people want to be re-assured, whereas truth is frequently not especially reassuring. It is an uncomfortable truth that because of human hubris, our species has treated nature as an expendable resource and ignored the desperate poverty of millions. Our failure to correct for our anthropocentric bias is partly because we are hypnotised by beliefs, narratives and abstract concepts, that just might, in the end, destroy us. These are not especially comforting thoughts.

For good or ill, an argument for mind in all things will seem necessarily a mystical concept to many. Strict materialists may see such philosophical ideas as an unwelcome attempt to shoehorn mystical and religious ideas into science. On the other hand, those with a more spiritual outlook may believe spiritual experience is the *only* way to reach this understanding.

I am not, in fact, advocating religious experience, or suggesting experience in general is more valid than reason as a path to knowledge. However, the philosophical starting point that there is ultimately no separation between the mental and physical, is part of many spiritual traditions, and there is a clear overlap between these traditions and the philosophical views of this book.

A commonly heard platitude is that all religions are saying the same thing, ultimately. This is a platitude when religion is taken as a whole, because faiths often define themselves by how they differ from other faiths. All major religions teach honesty, humility, compassion and tolerance as important moral qualities—as does any well-functioning secular society.

It is equally true that religious texts frequently contradict their own advice for tolerance and kindness, as most religions have sacred texts that anyone inclined to dogma can interpret as proving that theirs is the only true spiritual path, e.g. *Thou shalt have no other gods before me.* Through words, the world's religions frequently make incompatible claims and counter-claims for the supremacy of their belief system.

Yet, I would suggest that the mystics, be they Buddhist, Christian, Hindu, Jewish, Muslim, Sikh, or Sufi (to list just a few of humanity's many faiths) do usually speak the same essential truth. Mystics of all faiths take themselves out of society and devote themselves to spiritual practices, such as long periods of prayer or meditation. The purpose is to experience the peace and liberation that comes with the suspension of ordinary modes of thought, to gain understanding by breaking with the constant dialogues that for most of us occupy our stream of consciousness. This often leads to experiences in which there is a partial or even complete dissolution of the sense of self. Through mystical experience, there is often a feeling of liberation from personal history and its emotional baggage, and the Earthly self is recognised as a purely temporary form, that becomes less important than experiencing unification with the infinite.

Although theist mystics will relate to these experiences through God, their accounts sometimes appear to have more in common with the mystics of other faiths than with the more dogmatic members of their own faith. The mystic's knowledge usually comes from direct experience, where moments such as hearing birdsong, sunlight reflected in a drop of water on the side of a glass, or the sheer pleasure of basic physical movement becomes a source of joy and enlightenment. These worldly experiences become gateways for accessing the infinite, rather than our everyday experience of a world limited to its separate external forms.

In *After the Ecstasy, the Laundry,* the psychologist and Buddhist meditation teacher, Jack Kornfield, reports a Christian monk's description of such an experience:

> In the monastery garden I was doing a simple walking meditation back and forth... All of a sudden I was a two year old boy again taking his first step. It was glorious. Just the pleasure of putting my foot down, the spongy grass, the smell of the earth and the roses. All the plants and insects seemed much bigger, like when I was so young. It all felt so alive.[72]

After the Ecstasy, the Laundry draws together mystical accounts from people of all religious traditions, and focuses on the sometimes difficult task of integrating profound spiritual experience into daily life, after months or even years, of living in a spiritual community. What might be surprising to some, is that even after years of study and meditation, after profound mystical experiences, returning to secular life can still be challenging. Consumer culture generates considerable profit through making us feel we lack something, be it wealth, possessions, status or sex. Outside of spiritual communities, we all have different priorities, and the bills still have to be paid. Spiritual experiences can be tested by a society that is frequently structured to reinforce the wants, needs and insecurities of the egoic self.

For those of us who have not lived it, it is tempting to imagine anyone who experiences what is variously called *satori*, *nirvana*, unity with God, or even *enlightenment*, should then be above the petty emotional and material concerns of everyday life. We may think that the mystic gains some magical cloak of wisdom and perfection that shields them from pettiness for ever more. But the spiritual path rarely has a singular direction of travel. Although mystics may be profoundly changed by their experiences, they do not then inhabit bubbles of spiritual perfection from that moment on.

However, if the idea of connecting with the ultimate and having your individual identity dissolve away makes you uncomfortable, you are not alone! Even those who have chosen the spiritual path can fear losing their egoic self, a self that has been with us for as long as we can remember. Kornfield quotes a Sufi monk who, for a time, feared losing his identity as his spiritual practice progressed:

> As I looked at all I had held to be me, the separate individual, it began to unravel. At first there was an openness and emptiness, but with it came a rush of fear, a struggling to exist, some kind of terror... Only later when I

learned to let go into it, to let myself fall, did it open up into a cloudless sky where I disappeared.[73]

Consumerism is not the only reason our egoic self dominates our waking hours. From an evolutionary standpoint, our perception that we house a lasting self within gives us something to value, and motivates us to preserve our physical body. The majority of people believe that they have either a soul or some other lasting essence of themselves that goes beyond their physical form.

It is understandable that an experience which reveals the everyday perception of this carefully nurtured self to be ultimately empty or illusory could be a frightening, rather than a wholly elevating experience. Relating to *nothing* is not a particularly comforting or easy to grasp concept, so the human mind needs to make it into *something*.

Unless one is actively working to suspend the self through spiritual practice, for most of us, perceiving ourselves primarily as part of a greater whole for anything more than a moment will in any case, be a remote possibility. Our sense of who we are, is for most of us, created and maintained by an internal dialogue in the ego centre of the brain—you may be switching in and out of one of those dialogues right now as you read this book. The ego works out how to interpret our emotions and experiences, and views these as either enhancing, or threatening the self it seeks to uphold.

This abstract self has a purpose, because as we learn our specific strengths and limitations, the abstracted egoic self can help us prepare for the future, enabling us to put ourselves in situations where we can thrive. On the flip-side, when we become stressed, these internal dialogues can easily become self-destructive thought patterns, drawing heavily on regrets about the past or fears for the future, stoking anger and resentment at others, accompanied by an apparently endless stream of self-criticism.

The Harvard neuroanatomist Jill Bolte-Taylor was separated from her egoic self, not by spiritual practice, but through a massive and unexpected stroke in her left brain at the age of 37. Taylor recovered through eight years of patience and persistence, eventually regaining all the cognitive functions lost during her stroke. Although she recovered those functions, by rebuilding her brain, she became in some ways a different person, having been able to step back and observe the mechanics of her damaged brain from the inside.

At a ratio of 4:1 most strokes occur in the left hemisphere. In the book *My Stroke of Insight* Dr Taylor says the left brain interprets and produces language, occupies the past and future, and is concerned with detail and analysis.[74] For example, the left brain detects the edges of objects in three-dimensional space, defining objects as discrete and separate from other objects, and from our bodies. This means it perceives us as a solid, separate individual.

By contrast, she says, the right brain's perception is generalised, based in experience, occupying the present moment, and regarding us as a fully connected participant in the universe. To the right brain we are part of a greater whole, in which the individual is a fluid and expansive presence. What that means for our mental state is that the right brain is at peace, content with just being. By contrast, the left is always seeking something. Its role is to be on the look-out both for threats and opportunities. It is worth saying here, that although these hemispheric characterisations can vary, they apply to the majority of right-handed people, and 60% of left handed people. For the remainder of people, this characterisation (of the hemispheres) is usually reversed.

Within a few hours the stroke had created a golf-ball sized bleed in Taylor's left brain, largely knocking out language, memory and the ability to retrieve information. Forced out of her left brain and into her right, she became aware of the different priorities of the two hemispheres, as the internal egoic dialogue was shut-off for long periods at a time. For me, the description of her changed perception reads like an account of a deep mystical experience:

> I shifted from the doing-consciousness of my left brain to the being-consciousness of my right brain. I morphed from feeling small and isolated to feeling enormous and expansive, stopped thinking in language and shifted to taking new pictures of what was going on in the present moment. I was not capable of deliberating about past or future-related ideas because those cells were incapacitated. All I could perceive was right here, right now, and it was beautiful... Swathed in an enfolding sense of liberation and transformation, the essence of my consciousness shifted... I'm no authority, but I think the Buddhists would say I entered the mode of existence they call Nirvana.[75]

In this right-brain state she also became aware of the trillions of cells that compose and maintain her physical body, with a sense of wonder at their abilities, and gratitude for their hard work. There are perhaps similarities here with the philosophical approach to panpsychism of Charles Strong mentioned in Chapter 5—namely that our cells could be having many little moments of experience, and the fact that we seem to have a singular consciousness in our heads is because the sheer number of these cells makes it impossible for us to perceive their individual interactions.

In *My Stroke of Insight* Jill Bolte-Taylor wants us to have greater understanding of, and greater control over, which of the two hemispheres dominates our lives. She wants us to be aware of which programs we are nurturing by allowing them to run continually, and which we might set aside. It may seem obvious to state that our lives would be better if we recognised that the right-brain and left-brain are both essential to our existence and happiness, and that our culture inclines more to the *seeking story-telling* left-brained mode, than the *happy in the moment* right-brained mode. However, many of us will know how powerful the internal dialogue can be. It can dominate our consciousness with thoughts and analysis of emotional issues, in a way that does more to enforce an abstract sense of self than to provide any practical solutions to our problems.

Although this book is not in any way intended to be a self-help book, it is worth singling out a couple of practical points from *My Stroke of Insight*. First, no matter how dominant the story telling voice may appear to be, according to Jill Bolte-Taylor, the area it occupies in our brains is only the size of a *peanut*.[76] This noisy presence is, in terms of volume, a tiny portion of our 1.3kg brains.

Second, she has a pragmatic suggestion for how to co-exist with this story-telling self. We should acknowledge that it is trying to be helpful. It is, after all, monitoring our emotions and experience, comparing them to past experiences, and trying to interpret events in a way that will be useful to us in the future. As such, it needs to be heard, not suppressed, so she recommends setting aside a daily appointment in order to listen to this internal voice. If we honour the appointment, having gained its own special time, the egoic self is more likely to let us continue with our day. I have personally found this practice useful on occasion. If I find myself stuck in over-analysis in response to an event or an emotion, I assure the egoic self I will give it a voice later. Without judgement, I sit down, try to vocalise its concern so I can listen to its analysis at the appointed time. I will often then quietly ask myself, "will more analysis of this issue help the situation?" The honest answer is usually "probably not".

The value of the right-brained mode, where consciousness experiences the present moment, without it being filtered through the egoic self, has played a role in 20th century culture. As a counter to post-war conformity, for the Beats in the 1950s, and then the psychedelic movement of the 1960s, direct experience was sought in all its forms. Whether it was meditation, psychedelics, losing yourself in nature, the spiritual rituals of ancient cultures, or using more spontaneous methods to create art or writing, these were regarded as more authentic routes to knowledge than reason and analysis. They supposedly would take us to places everyday modes of thought could not.

Accounts from religious mystics, or experiences like those of Jill Bolte-Taylor are likely to be criticised as not being reliable descriptions of reality, because they are not sufficiently objective. But an insistence on objectivity taking precedence where human consciousness is involved can be a mistake, as there will be limits on what we can learn if subjective experience is downgraded. For example, Mark Solms discovered that dreaming and REM sleep are not the same thing, as most of his contemporaries believed. He made this discovery by simply *asking* patients known to have REM sleep whether or not they actually dreamed. Some of them did not. He also asked patients who did not have REM sleep whether they dreamed. Some did.

Solms was prepared to ask his patients about their own experiences and allow their words to influence his conclusions. He differed in this respect from many of his colleagues, who had been trained to be suspicious of the patient's own account of their experience, because objectivity was valued above all.

When I began writing this book, I knew that the idea of mind in all things was one that would appear more credible and natural to readers who are used to suspending their rational mind through meditation or other spiritual practices. These practices develop an understanding that left-brain modes of thought are only one part of our being. However this does not mean this knowledge is available only through mystical experience, or that there must necessarily therefore be a greater spiritual purpose to life.

In my early research I came across a couple of books, written by academics around 2010, with a similar metaphysical view to this book. One notable difference, however, was that they had an underlying belief in a greater spiritual purpose to humankind rediscovering its connection with nature and the spiritual.

When these books were written Occupy Wall Street protests had just followed the 2008 financial crisis, and with the Arab Spring young people attempted to overthrow corrupt, undemocratic governments in North Africa and the Middle East. There was a sense in both books that the human species was at a crossroads, a crossroads as much about our development as spiritual beings as the obvious practical risks in humanity exhausting its ecosystem. Both advocated a return to more spiritual practice in day-to-day life. This would not necessarily be through organised religion, but religious practices, meditation, community action, sometimes even supernatural experience. It was my personal impression that these authors were looking for signs that there was perhaps already an unstoppable momentum towards a better world.[77]

In the intervening decade, damage to the planet has worsened, many dictators have tightened their grip on power, and the gap between rich and poor has widened further. Sometimes I wish I could share the sense of purpose and meaning that those of faith have. People of faith are more likely to believe that, although the road may be rocky and difficult, there must ultimately be a greater guiding force, a path for us to follow, if we could just recognise it for what it is. Faith in God, or even just faith in the better nature of humans, is a faith that the intelligent human mind has an end-goal of being more compassionate and less egoic.

But we also know that the human mind is to an extent hard-wired to look for and find patterns and significance, even when these are lacking. We see faces in the clouds or in a window on a dark night, seeing things that are not there. Similarly, we are equally hard wired to look for patterns and an underlying order, (known as apophenia) which in extreme cases can plague the lives of people with severe mental illness, like schizophrenia.

Finding patterns can help us solve problems, and also give our lives a greater sense of meaning and purpose. We look for patterns when they are not there, and while this tendency has helped us develop technologically, given the constant stream of conspiracy theory garbage spread online, this tendency has not led humanity to a place of near perfection. For these reasons, I believe it is probably a mistake to look for a *why* to there being some aspect of mind in all things, and keep largely to the *how*.

9

The creative universe

Why do we need so many kinds of apples? Because there are so many folks. A person has a right to gratify his legitimate taste. If he wants twenty or forty kinds of apples for his personal use, running from Early Harvest to Roxbury Russet, he should be accorded the privilege... There is merit in variety itself. It provides more contact with life, and leads away from uniformity and monotony.
[Liberty Hyde Bailey, American horticulturist, *The Apple Tree, 1922.*]

The Bramley is one of the most popular cooking apples worldwide. The variety began with a single pip, planted in 1810 by a small girl in the cottage garden of a Nottinghamshire village. Some decades later the owner of the cottage, Matthew Bramley, gave permission for a local grower to cultivate the fruit on the condition it would bear his name. This mother tree of all Bramley apples started dying from a fungal infection in 2016. Yet the variety continues throughout the world thanks to grafting, a horticultural technique, possibly invented in China some 4000 years ago.

A pip from a Bramley apple will not grow into a new tree bearing Bramley apples, because its genetic code is different from the parent tree. In effect, every apple pip offers up a new variety of apple. Unless a branch from a specific variety gets grafted onto a neutral root stock, our favourite apples will quickly be lost. In a sense, grafting is a technique humans developed to restrain nature's inventiveness.

At the start of this book, I said the argument for there being some mind in all things is essentially one of subtraction —taking away ideas that do not work until it becomes the remaining option. But some positive case must also be made for what that means. In this final chapter I am going to speculate about the nature of this mind.

Since the Enlightenment, we have thought of nature as essentially a biological factory, that mechanically churns out similar copies of living forms. Darwinian evolution allows for the occasional happy accident to drive species change. But fruits like apples can change unpredictably with each generation. I see this as an example that, while the basic nature of the universe can be viewed as a mechanism, it also has a responsive and creative intelligence.

Intelligence and creativity have been an essential part of the human story. When hominid species started painting on cave walls and making decorative artefacts, the imperfections of the forms our ancestors created were not solely due to technical limitations. What on one level looks like a failure to correctly reproduce reality, was also a reflection of their priorities. For example, it is common in primitive art to find female forms with unnaturally large hips, which represented the ideals of being both well fed and fertile, in communities where neither of these could be guaranteed.

However, primitive humans also lacked the ability to produce high-quality brushes and paints, or the formal understanding of how the appearance of an object changes with perspective. In looking at early art, what should we attribute to the intent of the artist, and what to the limitations of their techniques and technology?

For the greater part of human history, attempts to be completely true-to-life would have been restricted by available technology. That was until the time of the Ancient Greeks. By 480BC, sculpture had developed enough for the human form to be represented faithfully in all three dimensions. Around this time, early Greek culture produced an exemplar of true-to-life representation, the Kritios Boy, which is a highly accurate human form in a relaxed, natural pose. Then something interesting happened to visual art after the Kritios Boy.

Having reached its representational peak, Ancient Greek sculpture did not spend the remaining centuries turning out perfected natural forms. Later Greek sculpture was often *less* true to life, *less* perfect. For example, athletes were often shown with opposing muscle groups in tension, something that is not possible in reality. Sculptors did this because, quite frankly, the end product just *looked* more interesting. Being entirely faithful to reality was, artistically speaking, a dead-end. Going beyond faithful representation to a heightened or altered reality, is more likely to engage and intrigue the mind.

Another example of this is the art movement of Impressionism. Given the continuing interest in Impressionist exhibitions, and how frequently Impressionist works are reproduced in our culture, this group of late-19th-century artists produced arguably the most popular art movement of all time. Yet when the Impressionists were working, photography was already offering another way to make images. Photography allowed for a highly accurate representation of any scene or object (albeit in monochrome), providing it remained still, and there was enough light to expose a photographic plate. Apart from adding colour, why bother with the effort of painting an imperfect scene from ordinary life, such as a woman and child crossing a field, some water lilies, or Parisians at a dance?

Earlier in the same century, one of the most forward thinking artists ever to have lived, J. M. W. Turner, had been both fascinated and dismayed by the possibilities of photography, fearing it would put artists like him out of work.

But Turner's fears were needless. We can shoot photographs with staggering image quality using our mobile phones, yet Turner's works, and the Impressionists' paintings continue to exert a pull on our consciousness to this day. What intrigues the viewer about Monet's water lily paintings, or Poppy Field at Argenteuil, is his mastery of visual ambiguity. How did the artist, using only a small number of paint marks, create a more emotive sense of place and time in the viewer's mind than a photograph could?

It is an ambiguity that goes beyond the analytical question of how the artist has achieved this effect. It touches on the realm of the experiential, where we might wonder "what am I looking at?", "am I a part of this scene?" Perhaps the lasting appeal of Impressionism is its reaching a sweet-spot in our consciousness, between representation and abstraction, finding the ideal balance between what can and cannot be resolved. Put simply, it could be the ambiguity in Impressionist paintings that gets our consciousness *going*.

Ambiguity occurs when the mind is presented with something it is not able to resolve, and the unresolved is also the defining characteristic of the quantum world. A quantum particle exists in a state of superposition where it can be in multiple states at once. It enters the world of classical physics when that superposition ends, giving it a definite location in time and space.

A degree of irresolution is also crucial to art, as its value is not determined by how well it reproduces a scene or an object. For a picture to go from mere representation to being a work of art, the image created must remain, in some way, unresolved in the mind of the viewer. Otherwise, we will likely lose interest.

Taking visual art as an example of how mind in all things works, then linking it with quantum physics will for many seem a rather tenuous connection. Bear in mind here, the parallel I am drawing—that the unresolved may be a key factor in both Impressionism and quantum physics, and that the unresolved is also key to awareness in living entities—should be taken as an observation, rather than an attempted proof of anything.

Yet, I believe it is worth considering the connection, because turning the unresolved state into a resolved state is perhaps the primary function of mind in all animal life. To return to the point of the first chapter, if there is an identifiable purpose to an entity having awareness, it is in enabling an organism to make decisions that help it to preserve its existence. At the same time, an organism needs to have sufficient curiosity about the unknown, or it would fail to adapt and take advantage of different circumstances.

Two of the defining characteristics of living things, homoeostasis and self-organisation, are there to resist change that would otherwise destroy an organism. Taking the most basic example of homoeostasis, as a mammal we have a fairly limited internal temperature range in which the body can safely operate. Consequently, we have evolved mechanisms like shivering and sweating that keep our temperature in a healthy range. All animals have specific abilities like these that allow them to self-regulate.

We also have self-organisation in that there are defined parts of our body and defined parts of our environment. For example, although our body is about 60% water, we do not dissolve down the drain in a rainstorm, because the mammalian body is organised enough to define and maintain the water in our body separate from its surroundings.

One recent scientific attempt to explain consciousness has been to view it as a defence against entropy. Entropy means that as time unfolds things have a tendency to break apart. For example, if you drop some dye into a glass of water, the dye will dissolve into the water, and not spontaneously reform itself into a single drop, inside or outside of the water. Then, if you drop the glass onto a hard surface and it breaks, it will not reform itself into a whole glass full of water. Although there is currently some debate in physics about how entropy functions when applied to black holes, everything else in the universe is considered subject to this sort of entropy.

The neuroscientist Karl Friston has taken a concept from thermodynamics, *free-energy*, to explain brain function. One reason Friston looks to free-energy to tackle the problem of consciousness, is that as a quantifiable physical energy, it can be applied to the brain function, and from this abstract models can be built that predict how brains will behave.

Entropy, without intervention, leads to an entity breaking apart and losing its coherent existence. In *The Hidden Spring*, Mark Solms argues that Friston's free-energy principle is key to understanding what consciousness is for. Consciousness in organisms may exist primarily to ward off entropy by minimising the level of uncertainty and unpredictability an organism encounters in its lifetime. It does this by having a model of the world it expects to encounter, a concept which is constantly updated by experience and perception, that tells the brain how closely its model matches the external world.

Minimising this unpredictability is important, as the more unpredictable a situation is, the more energy is required to process information to correct the internal model. In this way, Solms believes the mind works towards automatic and unconscious processing where possible, rather than having to direct the resources required by consciousness to process all sensory input.[78]

Where there is the potential for an unknown outcome, there is also the potential for entropy to increase, which could threaten the coherence of the organism. In this approach, Solms has found allies in both Karl Friston, and social neuroscientist Aikaterini Fotopoulou, who regard this minimising of unexpected states, and the *disambiguation* of the information received from the world around us, when it fails to match our internal mental models, as the brain's main purpose in consciousness.[79]

As we saw in Chapter 6, a growing view among neuroscientists is that the brain is not so much receiving information from external sources in order to observe the world. Rather, the brain is constantly making predictions about the world based on its internal model. It then uses the information received from the senses to error-check this internal model, make it more complete and up-to-date, more closely resembling the world we should anticipate in the future. The free-energy principle complements this neuroscientific view by saying a more complete and accurate internal model of the world minimises the energy that would otherwise have no specific direction or purpose. In this way it resists change. It resists entropy.

In all animals, the risks found in new and different circumstances must be balanced against the evolutionary need for curiosity. Animals risk exploring unfamiliar places and unfamiliar situations, because there are potentially great rewards in exploring the new. If there is a defining characteristic of consciousness in animals, and mind or awareness throughout the universe, it may be this balancing act of ambiguity, and disambiguation, the resolved and the unresolved. The key principle of living in an intelligent and aware universe would be that organisms are both constantly attracted to, and seeking to resolve, ambiguity.

I said in Chapter 1 that this argument for mind in all things is a reasoned argument that corresponds to the mystics' sense of the *aliveness* of all things—not only organic life controlled by brains full of neurons. Here, I am also suggesting we may live in a universe of intelligent creativeness, that reaches all the way down to the smallest scale.

Quantum computers have the potential to provide incredible processing power by harnessing ambiguity. Quantum processors have been with us since the early 2000s and work through the uncertainty of quantum superposition. To give an idea of their staggering potential power, consider that most home and work computers have 64 bits. Because they follow the laws of classical physics, each bit can have one state, either one or zero. With that in mind, we can perhaps imagine a far more powerful classical computer that has 500 bits. Now imagine this hypothetical classical computer swapping all of its 500 classical bits for 500 quantum bits, or qubits. It is predicted that the power of those 500 qubits is such that to build an equivalent classical computer would take more classical bits than there are atoms in the universe!

This phenomenal quantum power—the power of all the atoms in the universe—may not be purely theoretical because Google's Sycamore quantum computer had 54 qubits, and the Chinese Zuchongzhi quantum computer 66 qubits. In 2021 IBM announced their Eagle 127 qubit processor, followed in 2022 by the Osprey, 433 qubit processor. IBM's Condor, unveiled in December 2023, had 1,121 qubits. Over a 1000 bits is, theoretically, double the power of all the atoms in the universe.

The difference between a classical bit and a qubit is that where a classical bit can have only two states, one or zero, a quantum bit can have either of these states, both states, and all possible states in between, at the same time.

This fundamental difference means quantum computers are far from being drop-in replacements for the processors in our laptops and phones, or even the supercomputers at major institutions. They currently have only specific uses, and programming a quantum computer seems to be almost as much an art as it is a science. Quantum computers require significant error checking to be incorporated to compensate for a computing process that is based on probable, not predictable hardware. The art of quantum circuitry is in designing a series of quantum gates in a configuration that maximises the chances of getting a useful answer at the end.

Unlike a classical computer, quantum uncertainty means we cannot confidently say how a quantum processor has arrived at a result, and possible explanations are influenced by whichever interpretation of quantum physics one prefers. Taking the multiverse view, for example, a quantum circuit works by creating the conditions for quantum uncertainty, which causes the qubit to inhabit all possible states at once, exponentially increasing the potential computing power.

However, taking a view of the universe as a place of creative, aware intelligence, we would find a commonality between the behaviour of the qubit in a supercomputer, the sub-atomic particle in the double-slit experiment, the exciton in a leaf finding its way to the action centre, as well as an animal's consciousness balancing the rewards of the new with the uncertainty of creating "free-energy".

Living beings have an awareness of their environment, and will tend to react to their environment in ways that are largely, although not wholly, predictable. Unlike programmed machines, whose actions can be fully predicted if we know all the physical parameters, animals often exhibit behaviours that we might put down to mistakes on the one hand, or personality on the other—they surprise us by doing things that are not entirely predictable. One reason for this element of unpredictability may be that quantum processes are a part of the organic brain's operation.

Returning to the thought experiment of Chapter 4, we can make a similar observation about the quantum double-slit experiment. The result of this experiment is not entirely predictable, because although we can correctly predict whether an interference pattern will build up when single particles pass through one of two open slits, we cannot predict the position on the sensor where each *individual* particle will land. It may be that the possibility of another particle being close creates an unresolved situation, an aspect of free-energy, triggering the creativity of the universe, and so creating the interference pattern. The suggestion is that what links these, from the intelligent actions of living beings to the behaviour of particles, that what links the macro to the micro, is that there is a moment of uncertainty, of free-energy, which is then resolved.

For living beings, living with a certain level of unpredictability has an obvious evolutionary purpose. Through our extensive culture we sometimes abstract this into the arts like painting or music. Perhaps one reason these arts capture our attention is that they are triggering a reaction to what cannot be fully known, or resolved. My reason for linking visual art to quantum physics is that through that uncertainty and irresolution, we may also be said to be tapping into the power of an intelligent, behavioural, and creative universe.

Throughout this book, I frequently use the word "universe" while, in fact, I have actually only referred to the natural world here on Earth, and the building blocks of matter and energy. Panpsychist philosopher Rupert Sheldrake has observed that most people in the panpsychist camp do exactly what I have done, and limit their concept of mind to what is here on Earth, rather than attributing mind to entities out in space. While acknowledging the idea is somewhat speculative, Sheldrake thinks we should expect bodies like the Sun, stars and planets to have mind.

Physicist Gregory Matloff of City University of New York has argued that there could be a *proto-consciousness* field throughout the universe. He uses this to account for the observation that some smaller, cooler stars move more rapidly than larger ones, and fire off plasma in one direction, a phenomenon that cosmologists have yet to explain. As this creates a course change, it could be seen as the star exercising some sort of will, known as the volitional star hypothesis.[80] Matloff argues this happens in cooler stars because they are cold enough to have more complex chemistry than hotter ones. However, he also notes that an element of mind in a celestial body does not necessarily indicate an intelligence or awareness like ours. It could be, as Matloff points out, an aware intelligence only at the level of the slime mould. After all, whatever level of awareness the slime mould is at, it is sufficient to allow the slime mould to make a decision whether to move in one direction or another.

Do larger bodies have mind? Gaia theory treats the Earth as a self-regulating homoeostatic system, which may, for example, create plankton bloom to help shade the ocean and regulate temperature. True or not, unfortunately for the human race, Gaia self-regulation is unlikely to let us off the hook from our environmental damage. If we continue to disregard nature's warnings, its self-regulation might well mean getting rid of the one species that insists on throwing everything out of balance—our own.

It may be argued planets and stars are also homoeostatic systems, that respond to circumstances in ways that indicate a form of mind. As there is an action, that could qualify as a behaviour, a star releasing plasma has the *potential* for some intention or responsiveness in a way that a chair or rock cannot. However, the debate about whether there actually *is* mind in planets and stars is one to which I can add no useful insights. Knowing nothing about cosmology, I cannot say whether certain phenomena are caused by a form of intention or decision making out in the cosmos, and do not have the knowledge to weigh up alternative explanations.

The fact that not just animals, but also plants, and if we follow through on the thought experiment in Chapter 4, particles, respond to their circumstances in ways that look like a form of decision making does not mean experience *must* be present with that responsive behaviour. Taken with ideas such as the brainstem theory of consciousness, it does, however, raise the question of the breadth and depth of experience throughout the universe. Although I do not know whether the Earth has any awareness of itself as the entity *Earth*, I do think that in earthbound nature we can find some examples of creativity. Creativity, as observed previously, is also a known characteristic of intelligent animals such as humans.

I began this chapter with the variability of apples. That a fruit tree that needs to be grafted to maintain a stable variety, is for me, one example of nature's creativity. Most plants produce seeds that will lead to a new plant which is as well, or as badly, suited to its environment as its parent. Through natural selection, minor variations then crop-up that become part of the species' gene pool in time. Yet the reproduction of plant species like apples shows that nature can do more than simply churn out the same old product, generation after generation.

As fruits such as apples, are too large for birds to pick up, and then accidentally drop, the evolutionary strategy of wild apples is to entice larger animals to consume the fruit, then spread the pips to other locations in their droppings. A wider variety of fruit will be eaten by a wider variety of animals, increasing the range that a species can spread beyond its current location.

The genes of around 4 wild apple species are found in domesticated apples, and these domesticated apples show considerably more variety than the wild species.[81] Plant the pip from a Russet, a Cox's Orange Pippin or a Gala apple and you will not get a tree that produces mellow autumnal Russets, crisp Cox's, or soft-skinned, dappled red and yellow Galas. An apple tree could produce hundreds of apples per year for many decades, yet because the genetic makeup of the pip is different from that of the parent tree, none of the thousands of apples it produces in its lifetime will contain pips capable of growing the same variety again. This is why I have singled it out as an example of creativity in nature.

Pips may produce experimental, niche flavoured fruit that would only be enjoyable for a small number of people. As a result, orchards full of our favourite apples require human intervention, through the grafting of part of the original tree onto a neutral root stock.

With domesticated apples (and the other fruits we graft) rather than just repeating a theme, nature seems to be repeatedly throwing the dice. However, this extreme variability is not necessarily the best evolutionary strategy. If nature hit on a variety perfectly suited to its environment through this mechanism, the new variety would be lost in a generation if nature was left to its own devices. There is almost a sense that as we force nature to be more definite, nature responds by becoming more creative.

Creativity has also been part of human culture for tens of thousands of years, and in Chapter 2 I suggested we look for it in other animals too, for example, in whale and dolphin songs, which may have a purpose more like musical improvisation than our mostly utilitarian purposes for language.

Again, I am not claiming there is evidence that dolphins or whales are improvising like jazz musicians. Rather, I am asking that as large cetaceans have considerable capacity to communicate, what else might such species, with brains larger than our own, and a larger proportion of the brain than ours being cerebral cortex (supposedly the thinking reflective part of the mammalian brain) be doing with all that spare capacity? Their communication sometimes fulfils a utilitarian function, as they seem to communicate about the hunt, as they herd a school of fish together in tightly co-ordinated actions. However, as mentioned in Chapter 2, unlike ours, their large brains did not evolve a requirement to alter their environment in any significant way. What do we humans do with our brains when we are not dealing with immediate utilitarian needs? We are often creative. It is possible that analysis of dolphin and whale communication might reveal a creative intellect. A creativity that largely exists for its own sake.

Does a notion of creativity in other species anthropomorphise animals that are just products of evolution? Surely such creativity cannot fit with the ruthless impersonal mechanism of evolution?

I do not believe they are incompatible. Although the phrase that most commonly encapsulates our concept of evolution, *survival of the fittest*, suggests a mercilessly economical system, evolution frequently does more than the bare minimum required for each species to continue to the next generation. We have residual organs, (the appendix) and residual behaviours, such as the fight or flight mechanism, which is more of a hindrance than an asset to an airline pilot when alarms sound on the flight deck. If evolution were more efficient, we would lose these organs and behaviours quickly. But the products of evolution do not need to be the very best possible, only good enough to pass on genes to the next generation.

Nature also sometimes goes beyond what will ensure a species finds its place in the ecosystem, into something more experimental. Mammals are in general far more social than reptiles, for example. Clearly there are advantages to groups of animals forming tight social bonds, utilising co-operation for the prosperity and protection of the community as a whole. But unlike most physical attributes acquired through evolution, are the evolved behaviours that go with different animal communities really all that *necessary*? Could nature not be said to be experimenting with different minds, with different ways of structuring groups of animals, creating minds that have a strong emotional component, which are then greatly affected by positive or negative relationships within their community? Certainly, the social animal that grasps how its community functions and uses this knowledge to its advantage is more likely to thrive. But the rules that determine how these societies function, how an animal can best fit in with the inner lives of other members of the community, only exist because they too were created by nature.

One cruel but dramatic illustration of this was a study in the 1950s that tested the possibility of hope in rats. Wild rats were put in water deep enough to drown, and lasted only 15 minutes on average before succumbing. This was unexpected, because a previous group of domesticated rats of the same species had lasted much, much longer—about 60 *hours* on average.

The experiment was repeated with another group of wild rats, although this time, instead of letting them drown, after a few minutes they were rescued as they sank into the water. The rats were lifted out, held and dried. In very short order, the unfortunate rats were put back in the water. Second time around, these wild rats lasted a similar length of time to the domesticated ones, an average of around 60 hours.[82]

It seems for such a highly social animal the element of hope, hope that they would be rescued, had increased their chances of survival far more than a physical adaptation would have done. For example, even if the same species had developed webbed feet, this would not account for the stark difference in average survival time, from 15 minutes to 60 hours.

Taking the glass half-full view, we can view this as a reminder of the power of hope, which we might apply to our own lives. On the other hand, it also means the individual rat has an evolved dependency on its community and its perception of possible futures. It is a reminder that evolution is not only about a species adapting itself to the physical conditions in which it finds itself. The capabilities of many animals, including humans, can be heavily influenced by affective emotional states. These internal states can cause considerable problems as well as advantages, so why should evolution have us experience them?

Having just asked the *why* question, the religious view would be that a deity has given us a higher purpose, and arranged the world to operate in a way that we will never rationally comprehend. For those without faith, such as myself, explanations involving a benign deity conveniently ignore the apparent cruelty in nature.

But the secular alternative to a natural world created by an all-knowing, benign God is not necessarily that the universe is a random, unfeeling and unknowing place. I think there is an argument that although there is likely no *why*, no intent, no end-game that nature is working towards, there is an inbuilt creativeness, awareness and intelligence to our universe. This is seen throughout organic life, the operation of evolution, and the basic interactions of particles.

Seeing all of matter, all of nature and the physics that underlie them as a bit more creative, a bit more aware, a bit more *alive* than the current Western world view—inherited through centuries of industrialisation—may help us truly value the only known planet in the universe capable of giving our species a reasonably easy living. It may lead to real technological breakthroughs if we can harness nature's creativity and intelligence.

I hope that we will increasingly act to preserve nature, which is unlikely to come from any spiritual awakening, but because this is essential to our survival. Lasting environmental damage from climate change, species extinction and pollution, is accelerating and it is becoming apparent that our lifetimes will see more and more of these negative effects. The remaining question is how quickly will we get to a more healthy place, and how much damage will be done before our species finally balances our wants with our actual needs?

All biases are a barrier to knowledge and understanding, and there is an interesting psychological question raised by this book. Perhaps those of us who argue for an element of mind in all things have certain expectations of how the universe *should* be arranged? Perhaps people like me are only trying to unify mind and matter through an aware, behavioural universe because we are temperamentally disposed to look for a simple and elegant way to solve the mind/body problem. We may, in fact, be looking to put our round peg into a square hole. After all, a simple and elegant solution to a problem often looks like it *should* be the right answer, even when it's not.

The physical world can, in fact, be a rather inelegant place, with odd nooks and crannies. Take one of the building blocks of the standard model in physics, the masses of particles. I won't list them here, in their various units, but it is worth knowing that from the very small to the very large, there really is no neat formula or pattern that binds these masses together with a pleasing sense of unity. As we gain more accurate knowledge about the constituents of the physical world, the constants that modern physics is built on cannot be bundled neatly together with graceful mathematics, expressing some underlying order of beauty and simplicity. In fact, the foundations of the physical universe seem to lack any pattern. It looks far more like a collection of arbitrary values, numbers that just happen to be the way they are.

The question of *why* these masses are as they are can be answered with the strong anthropic principle. This says we must live in a sort of Goldilocks universe, because even a slight alteration to any one of those physical constants would have made the evolution of biological life impossible. The anthropic principle says the reason organic life exists at all is that we just happen to occupy a point in the history of the universe with the right cosmological pre-conditions for carbon to form, without which carbon based life would not exist. The anthropic principle's answer to the question of *why* from the last chapter is simple—cosmically speaking, we just lucked out. Following the anthropic principle, if the basic composition of the physical universe had been even marginally different, there would be no one here to ask the question *why*.

Maybe no unifying principle, from the macro to the micro, will ever be found, no matter how much the human mind likes to seek out patterns and rules. But it is not incompatible with the anthropic principle to view, for example, the unpredictability of the quantum world as an expression of nature's creativity, and to scale that up to the operation of our minds. I rather like panpsychist philosopher Philip Goff's description of quantum superposition as "a sort of refusal on the part of reality to be definitely one way or the other" (*Galileo's Error*, 2019). The idea that there is a form of creative intelligence at the base of the universe is not incompatible with objective science. It only admits the possibility that something we might call mind is present in all of nature.

One solution to the presence of mind, of will and intention, in living entities, is that there is some capacity for decision making, that expresses this creative intelligence, in all matter. To solve the problems created by a species that is so wrapped up in the abstractions of corporate profit, GDP, political and religious ideology, and our failure to repair our profoundly damaged planet, may or may not involve new equations at the frontiers of science. However, I believe the mystics' experiential insight, namely that there is an *aliveness* to all things, can be backed up by reason and some evidence. This should also help us build a more healthy relationship with a natural world we are entirely dependent upon.

Apart from a much needed humility about our place in nature, in practical terms, what would the idea of some aspect of mind in all things actually *do* for us? It is, after all, only another philosophical point of view. Although I have argued that it helps to plug the explanatory gap between unknowing matter and knowing beings, its main benefits may, in fact, not be in the study of human consciousness.

I do not anticipate the growing interest in the different approaches to the philosophy of panpsychism sparking a revolution in neuroscience, and metaphysical questions about consciousness will likely persist. Instead, I believe its value is more likely to be in moments of insight that lead to advances in the physical sciences, where currently the majority of matter is treated as wholly unknowing and unaware. Finding a way to mentally align ourselves more closely to the fundamental forces of nature, its intelligence and adaptability, through quantum computing, could help to rapidly test possible battery chemistries, or materials for cheaper and more efficient solar panels. The intelligence of nature, properly utilised, might give us the vital insights to more rapidly solve the problems we have created through industrialisation. More advanced engineering may turn out to be the greatest beneficiary of this philosophical view.

Is a different term needed for this philosophical view, given the word *panpsychism* has some negative connotations? Although it is not the most catchy pairing of words, if I were to suggest a label, it might be *behavioural materialism*.

The word materialism is retained as there is no reason to ditch the advances of good objective science. And I have been arguing that an element of mind in all things is visible in behaviours that suggest basic awareness, so behaviour is a term to retain. I say behaviour, not experience, because, as described earlier, we cannot directly attribute subjective experience to other entities, because communication about experience is currently only possible in our species. We can find neurological similarities with many other species that gives us an idea of their potential experience. At the next level down, there are mind-like behaviours—which is as much evidence as is likely to be available from everything else.

The simplest life forms, and aspects of the quantum world, exhibit behaviours that *may* include an aspect of awareness, but I believe it would be almost impossible to show that experience is also omnipresent. Each person's subjective experience remains their own. We assume others have an inner world similar to ours, because their brains are similar, and they behave as we would expect a similarly aware being to behave. Ultimately, though, without language, when it comes to attributing mind, making deductions from observed behaviour is our main source of evidence.

To conclude, as the European medieval mind gave way to the Enlightenment mind, humans removed the benevolent, malign, volatile and capricious minds from nature, and began to understand these natural forces as predictable, self-sustaining mechanisms. In times past, these mechanisms belonged to God. But post-Darwin, God was no longer required as the guarantor of change and order in the universe. At the same time, as our technology increasingly shaped the world for our purposes, humans grew more confident in believing we were in most respects, above, and in control of nature.

Throughout this book, I have continually used examples of the tendency to separate ourselves from other species, that came to dominate Western thought. However, now Darwin's continuity principle—that man differs from other species in extent rather than type—is more readily accepted by those working in the field of animal cognition, even if, as primatologist Frans de Waal wrote in 2016, the social scientists, philosophers, psychologists and anthropologists, are more likely to trail behind in this respect.[83]

In one of his notebooks, Darwin wrote:

Origin of man now proved.—Metaphysics must flourish.—
He who understands baboon would do more towards metaphysics than Locke.

This is an another guiding principle underlying this book: that an understanding of the nature of our own existence must include a proper understanding of the nature of other living beings' existence. This is not of secondary importance. Animal cognition is not just an interesting side-note to the extensive knowledge we have about ourselves. Instead, it is a crucial part of that understanding, because it challenges us to more fully account for concepts such as intelligence, experience, awareness and consciousness, what we mean by these terms, how they manifest and where they come from.

In the end, it could be that scientific materialism is right, that intelligent, aware experience is only possible with a specific arrangement of neural structures, as in the human brain. In addition, following the anthropic principle, the fact that these brains have come about is an accident of the arbitrary nature of our universe, which dictates that true consciousness is only possible in specific animals.

However, I cannot help thinking the occurrence of mind in a small group of species on one planet in the entire cosmos is rather a "something out of nothing" view, as it suggests an ability to start occupying a point of view that just appears out of nowhere. It also does not entirely fit with the no-brained slime mould's capacity to learn, remember, have likes and dislikes, weigh up options and make decisions—a fairly sophisticated set of behaviours that were long assumed to be dependent on a brain containing millions or billions of neurons.

An alternative to human minds being the only real points of self-awareness, popping up out of the darkness, is that human consciousness is one of many, many points of awareness in the universe, because the universe itself is, in a sense, a place of creativity and intelligence as much as it is a place of matter and energy.

Our lived experience is certainly one example of those points of awareness, which disappear almost as quickly as they appear. But with intelligent awareness found from the macro to the micro, our self-aware, reflective consciousness may also be just one form of awareness.

These points of view may occur because the entire fabric of the universe is constantly occupying such points of view. Some of these, like the animal mind, are easier to recognise and more defined because their behaviours are clear and obvious. Having lost some of our species' tendency to hubris, we are now finding more intelligent awareness throughout the animal kingdom.

I have, with some qualification, extended that behavioural tendency all the way down to the action of sub-atomic particles. I do not see this as adding unnecessary mystery to the no-nonsense matter-and-energy approach of materialist science. Rather, I regard the current materialist scientific view —that tiny moments of intelligence and sentience are popping up randomly to illuminate the darkness in a universe made almost entirely of unknowing and unaware matter—as a far greater mystery.

List of illustrations

Chapter 2.
Figure 1. Crow illustration by yatdzkr on Fiverr.com.

Chapter 4.
Figure 2. Thomas Young's sketch of light diffraction in double-slit experiment. Public domain.

Figures 3–5. By the author.

Chapter 6.
Figures 6 & 7. Edward Adelson's checker shadow illusion. Copyright Free Use. Copyright © Edward Adelson.

Chapter 7.
Figures 8 & 9. By the author, with icon sets from Culombio Art, da-vooda and GreenTana (istock.com) and UnifiArt (pixabay.com).

Index

Endnotes

Preface.

1. Neil DeGrasse Tyson. 2019. *Letters from an Astrophysicist.* Penguin. Page 108.

Chapter 1.

2. Heidel, Lawal, Felder et al. 'Phylogeny-wide analysis of social amoeba genomes highlights ancient origins for complex intercellular communication'. *Genome Research.* 2011. 21:1882-1891.

3. Goff, Philip; Seager, William; Allen-Hermanson, Sean, "Panpsychism", *The Stanford Encyclopedia of Philosophy (Spring 2024 Edition)*, Edward N. Zalta & Uri Nodelman (eds.). https://plato.stanford.edu/archives/spr2024/entries/panpsychism/.

4. End of the final episode, 'The illusion of reality'.

5. Kaminski, Pitsch, Tomasello. 'Dogs steal in the dark'. *Animal Cognition* (2013) 16:385-394

6. Tierney, John, 1991. 'Behind Monty Hall's Doors: Puzzle, Debate and Answer?' *New York Times*. July 21st.

7. Herbranson WT, Schroeder J. 'Are birds smarter than mathematicians? Pigeons (Columba livia) perform optimally on a version of the Monty Hall Dilemma'. *Journal of Comparative Psychology*. (Feb 2010) 124(1):1-13.

Chapter 2.

8. For brain to spinal column ratios in mammals other than dolphins: Warden, C. J. 'Animal intelligence'. *Scientific American*. June 1951. 185, 64.

    For brain to spinal column ratios in three dolphin species: Ridgway, Flanigan, McCormick. 'Brain-spinal cord ratios in porpoises: Possible correlations with intelligence and ecology'. 1966. *Psychonomic Science*. 6(11), 491–492.

9. See Harvard website, bionumbers.hms.harvard.edu. In the pdf file named 'Brain weight, encephalisation quotient and number of cortical neurons in selected mammals.pdf' the EQ ratings are 7.6 for humans, 5.3 for the bottlenose dolphin, and the highest scoring primates get an eq of 4.8.

10. C.S. Lewis. 1984. *The Discarded Image: An Introduction to Medieval and Renaissance Literature*. Cambridge University Press. Page 92.

11. 'Animal Intelligence'. *Nature*. 1882-09-07. Volume 26, Issue 671.

12. Robert L. Pitman, the marine ecologist who witnessed the incident co-authored an article in *Natural History magazine* in November 2009, with John W. Durban, titled 'Save the Seal! Whales act instinctively to save seals'. https://www.naturalhistorymag.com/exploring-science-and-nature/131929/save-the-seal. (See also https://www.science.org/content/article/why-did-humpback-whale-just-save-seals-life). Although the explanations here are in the parsimonious category, and do not include intentional altruism, the article does conclude: "When a human protects an imperilled individual of another species, we call it compassion. If a humpback whale does so, we call it instinct. But sometimes the distinction isn't all that clear."

13. Galit Shohat-Ophir, Karla R. Kaun, Reza Azanchi, Hany Khammis Mohammed and Ulrike Heberlein. 'Sexual Deprivation Increases Ethanol Intake in Drosophila'. *Science*. 2012. Vol: 335, 1351–1355.

14. In her book *Through a window. My Thirty Years with the chimpanzees of Gombe*. 2010, (pub. Mariner Books classics) Dr Goodall recognises cruelty in chimpanzee behaviour, although in Chapter 10, *War,* she does seem to believe there is a difference of intentionality with human cruelty compared to chimpanzees. I do not know if Dr Goodall sees this as a difference of type or extent, but as we know chimpanzees can plan future actions, it would seem possible for chimpanzees to intend to be cruel in the future, as we do.

15. Liangtang Chang, Shikun Zhang, Mu-ming Poo, Neng Gong. 'Spontaneous expression of mirror self-recognition in monkeys after learning precise visual-proprioceptive association for mirror images'. *The Proceedings of the National Academy of Sciences* (PNAS). 2017. Vol: 114 (12) 3258–3263.

16. Domestic cats may deserve a pass. A series of clips on social media around 2020 showed different young people holding their cats as they filmed themselves with their phones. When the cat looks towards the camera and screen, the owners tap the screen, and the phone imposes a cartoon cat face onto the owner's face. Several of the cats notice the change on screen, immediately look concerned and turn to look *behind.* The screen image is effectively working as a mirror, and it seems to me many of these cats know they are looking at an image of themselves and their owners. It is just possible these videos were faked—perhaps someone else makes a noise behind the owner when the owner's face changes. But this seems unlikely, because the video was just goofy social media click-bait, not one claiming that cats have self-recognition. I believe it would be fairly easy to reproduce this procedure with scientific controls.

17. Erica F. Andrews, Raluca Pascalau, Alexandra Horowitz, Gillian M. Lawrence and Philippa J. Johnson. 'Extensive Connections of the Canine Olfactory Pathway Revealed by Tractography and Dissection'. *Journal of Neuroscience*. August 2022. 42 (33) 6392–6407.

18. See the website of The Wild Dolphin research project, CHAT section (Cetacean Hearing and Telemetry). https://www.wilddolphinproject.org/our-research/chat-research/

19. Yamamoto, Chisato & Kashiwagi, Nobuyuki & Otsuka, Mika & Sakai, Mai & Tomonaga, Masaki. 'Cooperation in bottlenose dolphins: bidirectional coordination in a rope-pulling task'. Oct 2019. *PeerJ*. 7. e7826. A free full text pdf can be found on www.researchgate.net.

20. Schrödinger's cat, locks an imaginary cat into a box, with a 50/50 chance of it being killed by a poison triggered by uncertain radioactive decay. It was written by Schrödinger in 1935 in a paper called 'The present situation in quantum mechanics' to highlight problems with quantum theory of the time.

   Note that in Carlo Rovelli's 2021 book *Helgoland*, his description of Schrödinger's cat has the cat exposed, or not exposed, to a sleeping draft, instead of a deadly poison. I'm sure Dr Rovelli is aware of Schrödinger's original phrasing. For me this rewording shows that even when describing a decades old thought experiment, today's scientists have an eye on negative public perceptions of animal experiments in science.

Chapter 3.

21. See TEDxSeattle 2017 talk by Suzanne Simard, *Nature's Internet: How Trees Talk to Each Other in a Healthy Forest*.

22. For more on early experiments where music was tested as a growth stimulant, with poor controls, see Chapter 4 'What a plant hears' from Daniel Chamovitz's 2012 book *What a plant knows. A field guide to the senses*.

23. Brenner ED, Stahlberg R, Mancuso S, Vivanco J, Baluska F, Van Volkenburgh E. 'Plant neurobiology: an integrated view of plant signalling'. *Trends Plant Sci.* 2006. 11(8):413–9.

24. Myers, Natasha. 'Conversations on Plant Sensing: Notes From the Field'. *Nature and Culture.* 2015. 3:35–66.

25. Gagliano, M., Renton, M., Depczynski, M., & Mancuso, S. 'Experience teaches plants to learn faster and forget slower in environments where it matters'. *Oecologia.* 2014. 175(1). 63–72.

26. Dussutour, A., Latty, T., Beekman, M., & Simpson, S. J. 'Amoeboid organism solves complex nutritional challenges'. *Proceedings of the National Academy of Sciences of the United States of America.* 2010. 107(10), 4607–4611.

27. Greenwood, Veronique. 'Sleep Evolved Before Brains. Hydras Are Living Proof'. *Quanta Magazine.* May 2021. www.quantamagazine.org/sleep-evolved-before-brains-hydras-are-living-proof-20210518

28. TEDxJaffa talk by Ariel Novoplansky. *Learning Plant Learning.* 2012.

Chapter 4.

29. Einstein's use of the term "spooky" is usually taken as his disbelief that particles can move in perfect synchronicity across any distance, implied by quantum entanglement.

However, theoretical physicist Sabine Hossenfelder has a different take on Einstein's meaning in her YouTube video, *What did Einstein mean by Spooky Action at a Distance?* www.youtube.com/watch?v=Dl6DyYqPKME. Re-visiting two instances when Einstein used this phrase (in 1927 and 1948) she believes Einstein was concerned that as a particle is measured, and its location goes from being probable to definite, it might imply some *knowing* between the particle on one side of a screen and the wave function on the other. She also believes too many people dismiss Einstein as having this key concept in quantum mechanics wrong—and that, instead, his concern was that quantum mechanics was incomplete.

30. In 1961 Claus Jönsson of Tübingen University performed the first quantum double-slit experiment with electrons. In 1989 Hitachi labs did the first with single electrons. For a summary of the history of the quantum double-slit experiment see *The Institute of Physics* website physicsworld.com, "The double-slit experiment". 2002. Sep 1. p15. This online version has been revised and updated to include more recent findings. physicsworld.com/a/the-double-slit-experiment/

31. Yoon-Ho Kim, Rong Yu, Sergei P. Kulik, Yanhua Shih, and Marlan O. Scully. 'Delayed "Choice" Quantum Eraser'. *Physical Review Letters.* January 2000. 84, 1–3. The use of quotes around the word "Choice" is the wording of the paper's title.

32. See Dr Jim Al-Khalili's 2013 Royal Institution lecture for an entertaining description of the measurement problem. *Double Slit Experiment explained!* www.youtube.com/watch?v=A9tKncAdlHQ. Here Dr Al-Khalili shows how bizarre the measurement problem really is (from a strict materialist perspective).

He humorously imagines the atoms being warned that they are being watched, before the scientists sneakily turn off the measurement device—and yet somehow the atoms seem to know about this subterfuge! Theories to explain the measurement problem, such as the pilot-wave theory, and the multiverse, have gone in and out of favour. Most importantly, the measurement problem is far from being solved, and in the video above, Dr Al-Khalili says that if someone can find the answer they will get a Nobel prize, which is still to be claimed. To be clear, I have *not* referenced Dr Al-Khalili's talk to suggest he is endorsing a panpsychic interpretation here. Rather it shows that for materialist science, an aspect of mind in matter sounds like *this way madness lies!*

33. Tononi Giulio and Koch Christof. 'Consciousness: here, there and everywhere?'. Phil. Trans. R. Soc. 2015. B37020140167. See Section 5, where Koch and Christof recognise parallels between IIT and panpsychism.

34. Some common criticisms of the Simulation hypothesis. First, if we are in a simulation, someone or something must have created that simulation, yet we are the only beings we know are capable of creating any simulation, convincing or unconvincing—claims that whoever or whatever created the simulation would not allow us to know about our being in a simulation is the type of get-out clause used by flat-earthers and conspiracy theorists.

Second, although our technology can be said to create many sophisticated computer simulations, which from the outside look and behave more and more like the real world, we have no evidence that any entities in those simulated realities are actually having an experience, something we know each of us is capable of. Third, for physicists, multiple simulations would violate calculated limits on the total amount of information in the universe. These are several other issues with the simulation hypothesis.

35. Charles W. Misner, Kip S. Thorne, John Archibald Wheeler. 1973. *Gravitation,* Princeton University press. Page 1217.

36. See https://www.youtube.com/watch?v=r8u8wjJ-jZg *From enigmas in physics to a structural version of idealism.*

37. Franco MI, Turin L, Mershin A, Skoulakis EM. 'Molecular vibration-sensing component in Drosophila melanogaster olfaction'. *Procedures of the National Academy of Sciences U S A.* 2011. 108: 3797-3802.

    A similar effect found in humans, with types of musk: Gane S, Georganakis D, Maniati K, Vamvakias M, Ragoussis N, et al. 'Molecular Vibration-Sensing Component in Human Olfaction'. *PLOS ONE.* 2013. 8(1): e55780.

38. For an explanation of the "quantum beat" see Jim Al-Khalili and Johnjoe McFadden. *Life on the Edge. The coming age of quantum biology.* 2014. Penguin/Random House. Page 173.

Chapter 5.

39. David Skrbina, *Panpsychism in the West*, MIT Press, 2005.

40. Martin Luther wrote an introduction to a 1542 satirical book by Alber Erasmus, *The Barefoot Monks' Jester*, where Luther claimed St Francis' stigmata came from an injury caused by a roasting spit, during a fight with his friend St. Dominic.

41. Julian Baggini. 2018. *How the World Thinks.* Granta. Page 32.

42. *The Acquaintance Hypothesis* is described by Conee, E. 'Phenomenal Knowledge'. *Australasian Journal of Philosophy.* 1994. 72: 136-150.

43. I say this, knowing that currently the quark is considered the smallest unit of matter. However, the quark is a recent discovery, and as physics is not yet done understanding the universe, it seems likely other smaller building blocks to physical reality will be discovered. Will we ever be able to confidently assert that we have reached the smallest possible unit of matter?

Chapter 6.

44. Mark Solms. 2022. *The Hidden Spring. A Journey to the Source of Consciousness,* Profile books. Page 142.

45. Erin J. Wamsley, PhD and others. 'Cognitive Replay of Visuomotor Learning at Sleep Onset: Temporal Dynamics and Relationship to Task Performance'. *Sleep.* 2010. Volume 33, Issue 1, January. Pages 59–68

46. A summary of research into birdsong, learning and role of dreaming. Margoliash, D., & Schmidt, M. F. 'Sleep, off-line processing, and vocal learning'. *Brain and language.* 2010. 115(1), 45–58.

47. For a summary of studies involving Matthew A Wilson see https://news.mit.edu/2002/dreams. Here are two examples.

During slow wave, non-REM sleep, the daytime activity patterns in rats were found replayed at about 20x real time speed: Lee, A.K. & Wilson, M.A. 'Memory of sequential experience in the hippocampus during slow wave sleep'. *Neuron.* 2002. 36. 1183–1194. In REM sleep a similar effect was found, but at real time: Ji, D., Wilson, M. 'Coordinated memory replay in the visual cortex and hippocampus during sleep'. *Natural Neuroscience.* 2007. 10, 100–107. A free pdf of this study is also available on researchgate.net.

48. See Wikipedia,
https://en.wikipedia.org/wiki/Cerebral_cortex#Structure

49. Mark Solms. 2022. *The Hidden Spring. A Journey to the Source of Consciousness,* Profile books. Page 60.

50. I should add here that Mark Solms is philosophically not a panpsychist. In fact, in *The Hidden Spring* he only mentions panpsychism briefly, and describes it as "worrying".

51. See https://uni-tuebingen.de/ news page on 24/09/20, titled: 'Tübingen researchers show conscious processes in birds' brains for the first time'.

52. A summary of Higher Order theories of consciousness, can be found here.
https://plato.stanford.edu/entries/consciousness-higher/

53. Dolphins are classified as even-toed ungulates, because they share a common ancestor with modern herd animals, from 60 million years ago.

54. Morrison R, Reiss D. 'Precocious development of self-awareness in dolphins.' *PLOS ONE.* 2018. 13(1): e0189813

55. Moshe Shay Ben-Haim et al. 'Disentangling perceptual awareness from non-conscious processing in rhesus monkeys (Macaca mulatta)'. *PNAS.* 2021. Vol. 118 (15). March 30. e2017543118. This article also describes the same study. https://phys.org/news/2021-03-monkeys-visual-world-people.html

56. Briefly, I want to give my personal view on free-will. It's a debate that divides people strongly into pro or anti camps, so feel free to ignore these points. I think that the idea that its existence can be proven or refuted largely ignores the fact that free-will is created by its context. As such I do not believe it is something that can be *proven* one way or the other. It is a necessary social, legal and moral construct, which therefore cannot be proven into existence, or disproven out of existence by science.

Take a far less consequential concept. Imagine if researchers claimed EEGs have proven that ultimately there is no such thing as *aesthetics* or *good taste*? We know there are some tried and tested principles of aesthetics, such as the golden ratio, or the rule of thirds in composition which make a photograph or painting work artistically. But we also know good taste and aesthetics are in the end subjective. If a group of scientists claimed to have proven there's no such thing as good taste they would be largely ignored, because the vast majority of people believe these concepts have no objective reality anyway.

Of course, free-will has a far greater significance in our lives. Maintaining a balance between personal responsibility for our actions, and actions which might look freely chosen, but are primarily the result of severe mental illness, brain damage, early childhood trauma, etc, is essential to a well-functioning and just society. However, there are influential groups who believe free-will is essentially absolute. In the US in particular Christian evangelical groups have more influence on government, academia and society than in most of Europe. The concept of free-will is bound up with the concept of a soul that will one day be subject to divine judgement. I wonder if the reach of the American evangelical movement is a reason some scientists feel the need to refute free-will? This seems especially relevant when more extreme groups believe free-will justifies psychologically harmful practices like "gay conversion therapy" which is wrongly based on the idea that one's sexuality is a personal moral choice.

57. See the subheading, *10 signs of consciousness* in 'Why be conscious?'. *New Scientist.* 2017. Volume 234, Issue 3125, 2017, p28–31.

58. See this brief interview with Dr. Bruno van Swinderen (Queensland Brain Institute) https://www.youtube.com/watch?v=lc8UbqKO01Q

59. See, www.npr.org/transcripts/654730916

60. M Kohda, R Bshary, N Kubo and others. 'Cleaner fish recognize self in a mirror via self-face recognition like humans'. *PNAS.* 2023. Volume 120. No. 7 February 14. A good summary of this study is also here, https://phys.org/news/2023-02-bluestreak-cleaner-wrasse-mirror.html

Chapter 7.

61. In 2022 panpsychic philosopher Dr Philip Goff edited a book called *Is Consciousness Everywhere? Essays on panpsychism.* Imprint Academic. This is an intriguing collection of essays about panpsychism from scientists, philosophers, and theologians, which arose from Dr Goff's own 2020 book, Galileo's Error. In the first edition, the first essay by physicist Sean Carroll gives a well considered "no" to panpsychism in science.

But in my opinion, in the opening paragraph Dr Carroll does fall prey to the myth of human exceptionalism, when he gives the now incorrect figure of 100 billion neurons in the human brain.

62. Philip Rice and Patricia Waugh. 1989. *Modern Literary Theory. A reader.* Edward Arnold. Page 5.

63. Although English has lost its gendered nouns, it does not follow that English speaking societies are more cruel to animals than others. Rather, justifications for animal cruelty may come with a different rationale. Compare for example the justifications given for traditional "sports" of fox hunting in England and bull fighting in Spain. While both will cite tradition in their defence, defenders of fox hunting will often claim it is a necessary measure for pest control. Whereas in Spain, bull fighting would rather be defended as giving the bull a chance to express its true nature through the fight. Perhaps such a difference reflects the continuing use of gendered nouns in romance languages?

64. Philosopher Derrida attained celebrity status in a way that was perhaps only possible in France. While visiting a civil rights group in Prague, the communist Czech authorities imprisoned him on trumped up charges. The Mitterrand government negotiated his release, and the philosopher was rather overwhelmed by the huge crowd that greeted him on his arrival at the Gare du Nord in January 1982. In 1983 his notoriety meant he wrote for and appeared in a (critically panned) independent film, *Ghost Dance*.

65. For example, the Step Function is discussed in episode 1202, 'How do Humans Differ from Other Animals?', in the PBS documentary series *Closer to Truth*.

66. Suzana Herculano-Houzel. *The Human Advantage: How Our Brains Became Remarkable.* MIT Press. Page 22.

67. See Chapter 6 *White Dwarfs and Little Green Men* of Sagan's 1979 book *Broca's Brain.*

68. See the TedXVictoria talk by Genevieve Von Petzinger, The Roots of Religion. In the early part of this talk she mentions a theory of enhanced working memory, devised by archaeologist Thomas Wynn and neuropsychologist Frederick Coolidge, as a possible start for religious thinking in humans. I'll let the reader search for this talk as I have not been able to find a stable link.

Chapter 8.

69. I confess to never having read Lamarck's *Philosophie Zoologique*. I suspect that is also the case for most of the people who now claim Lamarck was right and Darwin wrong! It is not an easy read, and my understanding of Lamarck comes mostly from Hugh Elliot's informative 70 page introduction to his translation of the text, (Hafner, 1963). While Lamarck's work had many flaws, like the very best scientists, his aim was to advance knowledge and truth. For this, he gave up a privileged background, and ended up buried in a paupers' grave. We cannot now pay him our respects, as his only surviving daughter could not afford any monument. Like so many forgotten people who died in poverty, his remains were likely scattered at Montparnasse cemetery, around five years after burial.

70. The concept of gemmules is also known as Darwinian Pangenesis.

71. Plank et al. 'Prenatally traumatized mice reveal hippocampal methylation and expression changes of the stress related genes Crhr1 and Fkbp5'. *Translational Psychiatry*. 2021. 11:183. About 15 pregnant mice were subjected to a random 2-second electric shock.

They and their offspring were compared to a control group who did not get the shock. I have commented on the ethics of such experiments in the penultimate endnote for Chapter 9.

72. Jack Kornfield. 2000. *After The Ecstasy, The Laundry How The Heart Grows Wise On The Spiritual Path.* Rider. Page 30.

73. As above. Page 57.

74. See also Dr Bolte-Taylor's 2008 video on the Ted channel, *My Stroke of Insight* for her humorous and then moving account of her experience. This is my personal favourite of all Ted talks, and Dr Bolte-Taylor's book is a fascinating and detailed account of her experiences. It also has some important thoughts about aiding the process of recovery from stroke.

75. Jill Bolte-Taylor. 2008. *My Stroke of Insight. A brain scientist's personal journey.* Yellow Kite, Hachette. Page 49.

76. The area of the brain Dr Bolte-Taylor refers to as "peanut sized" is part of the language centre of the higher cortex.

77. In my reading of these two books, I had the impression their authors believed in a higher spiritual purpose to, and destination for, the current human condition. The first was *The Coming Interspiritual Age* by Kurt Johnson and David Robert Ord, (Namaste Publishing, 2013). Amongst other things, this took what was then the recent Arab Spring and Occupy Wall Street protests as potentially part of a more profound shift in human nature towards freedom and equality. The second was *The Living Universe: Where Are We? Who Are We? Where Are We Going?* Duane Elgin, Deepak Chopra. (BK Life, 2009). Elgin expected that the internet would lead to greater understanding, and reduce the likelihood of violence and exploitation as the world can bear witness to wrongdoing anywhere in the world.

In my opinion, the decade that has followed these two books does not, unfortunately, suggest humanity is reaching towards a higher spiritual purpose. I sincerely wish that we were.

Chapter 9.

78. See section 7, *A surprise*. Solms, Mark. 'The Conscious Id'. *Neuropsychoanalysis*. 2013. Volume 15. Issue 1, p5–19.

79. See commentaries to 'The Conscious Id' (in the above) by Fotopoulou (page 35) and Friston (page 38).

80. Greg Matloff. 2020. *Starlight Starbright: Are Stars Conscious?* Curtis press.

81. Spengler R. N. 'Origins of the Apple: The Role of Megafaunal Mutualism in the Domestication of Malus and Rosaceous Trees'. *Frontiers in plant science.* 2019. 10, 617. https://doi.org/10.3389/fpls.2019.00617

82. Richter, C. P. 'On the phenomenon of sudden death in animals and man'. *Psychosomatic Medicine*, 1957. 19. p191–198.

There are many descriptions of this rather infamous experiment available online. If you can stand it, you can read Dr Richter's own description of how they found the optimum temperature for the length of time a domesticated rat will swim. Also, how they trimmed the whiskers of all the rats studied, which had something to do with preventing contamination (I wasn't clear why). Some rats were also hooked up to EKG machines while in the water, and found to die from a slowing rather than rapid heart rate, which is one reason for concluding hopelessness, not panic, was the final cause of death.

In this book, I have described several scientific experiments that I do not condone, and mostly side-stepped the issue of animal cruelty in science, as this subject deserves more space than is available here, and deserves the opinions of those better qualified than I. While the Richter experiment is symptomatic of its time, it was important to include, because its stark conclusion has made a significant contribution to animal and human psychology. It struck me that the DNA methylation experiment with mice in Chapter 8, where pregnant mice received a two second shock to induce significant stress, is perhaps little better, and is permitted in modern universities.

The Richter experiment was conducted in the 1950s, and it would seem hard to justify it being repeated in the 21st century. I have since discovered that the University of Bristol, in my home city, is one of a small number of institutions permitted to carry out a similar 'forced swim test' with rats and mice. According to both the Telegraph and Times in March 2023, it was conducted 189 times during 2022 by Bristol University. According to the University of Bristol website in July 2023, it runs for a maximum of 6 mins for rats, 5 mins for mice, before they are rescued, euthanised, and tissue samples taken.

Supposedly, the experiment in antidepressant development. However, few pharmaceutical companies use the forced swim test now, and this study concluded it has little to no value in antidepressant development. Sewell F, Waterson I, Jones D, Tricklebank MD, Ragan I. 'Preclinical screening for antidepressant activity - shifting focus away from the Forced Swim Test to the use of translational biomarkers'. *Regul Toxicol Pharmacol*. 2021. Oct. 125:105002.

83. Frans De Waal. 2017. *Are we Smart enough to know how smart animals are?* Granta.

Written by Matthew Benton

Matthew is an IT professional from Bristol UK with a BA in English and Philosophy. Brought up to be equally passionate about the arts and sciences, he has also always been fascinated with how other animals perceive the world. He meditates, walks, and is an enthusiastic tennis player. He has been fortunate enough to swim with wild bottlenose dolphins on several occasions. *The Case for a Living Universe* is his first book.

Edited by James Kingsland

James has been in science journalism and publishing for over 25 years. He worked as senior sub-editor at the UK's New Scientist magazine, and was science production editor at The Guardian newspaper. James has written two books: *Siddhartha's Brain: Unlocking the Ancient Science of Enlightenment* (2016) Mariner Books. And *Am I Dreaming?: The New Science of Consciousness, and How Altered States Reboot the Brain* (2019), Atlantic Books.